# 50 Simple Things You Can Do To Raise a Child Who Loves Science & Nature

Marion A. Brisk, Ph.D.



Macmillan • USA

**Macmillan • USA**
A Simon & Schuster Macmillan Company
1633 Broadway
New York, NY 10019-6785

Macmillan Publishing books may be purchased for business or sales promotional use. For information please write: Special Markets Department, Macmillan Publishing USA, 1633 Broadway, New York, NY 10019.



Manufactured in the United States of America
10 9 8 7 6 5 4 3 2 1

Library of Congress Number: 97-071526
ISBN: 0-02-861935-8

Book design by Rachael McBrearty—Madhouse Productions

**An Arco Book**
ARCO and colophon is a registered trademark of Simon & Schuster Inc.
MACMILLAN and colophon is a registered trademark of Macmillan, Inc.

# Contents

## Section 2 Kitchen Science

### Icecapades



### Acids and Bases



### Water, Water, Everywhere



### The Toys



## Section 3 Learning About Nature and the Environment

## Section 4 Games and Puzzles

### Guessing Games



### The Puzzle Place



### The Games Children Play



## Section 5 Places to Go, Things to Do

# Introduction

## To the Parents

You are your child's first and most important teacher. The purpose of this book is to provide you with simple, straightforward activities and experiments that you can share with your child to encourage a genuine interest in science and nature. With just everyday household items, you can spend many delightful hours supporting the curiosity and sense of wonder that children have about the natural world.

Today, more than ever, it is important that our children develop the ability to understand science and technology. As scientific breakthroughs continue to change the way we live, our children will need scientific knowledge in order to make informed decisions. Just compare a newspaper of 20 years ago to one of today to see how science and technology have become part of our lives. It is also particularly critical that our children develop an appreciation and understanding of the life-support systems of our environment as we strive to decrease pollution, slow down global warming, and end our assault on the other species of the planet in the twenty-first century.

All children are natural scientists; they are curious about the world around them, observant about what they see, and creative whenever they are allowed to be. When we spend time doing experiments with our children, or we go outside exploring nature together, we are nurturing their curiosity, encouraging their innate interest and enthusiasm, and fostering an appreciation of the natural world. Some activities that you do with your child may even develop into a lifelong interest: Rock hunting may give birth to a geologist, or making a bird feeder may spawn an avid birder who helps to slow the rapid decline of many bird species around the world. You are opening new doors for your child every time you explore another topic together. Girls in particular need to receive positive messages about science and math early on, and continued encouragement to help counteract the many negative influences—from schools, the media, and often friends—that can erode and destroy a desire to learn and participate in the world of science.

The activities and experiments that follow are divided into five sections: **Shapes, Colors, and Light; Kitchen Science; Learning About Nature and the Environment; Games and Puzzles;** and **Places to Go, Things to Do.** Activities and experiments were chosen because they tend to be fun for children (and, in many cases, for adults as well), such as bird watching, rock hunting, making solutions change color, or experimenting with plants. They were chosen also because they demonstrate some important scientific facts and concepts. These early experiences will be helpful later on, even after the elementary years.

Within sections some activities and experiments are grouped together because of a common theme. Each exercise is divided into **The Main Idea, What You'll Need, How Long It Takes, What You'll Do,** and in some cases, **Digging Deeper, Tips,** and **References.** Digging Deeper sections generally involve activities or experiments that older children in particular can understand; tips are provided where further information will help make the project more rewarding; and references and Web sites are given for additional study.

Many of the experiments and activities will help your child develop an appreciation and respect for the environment. It is vital today for all of us to recognize our dependence on the Earth's water, air, land, and living world and learn how to preserve them. With about 100 animal and plant species disappearing from the planet daily, global warming from consumption of fossil fuels portending climate changes, worldwide soil erosion threatening food supplies, and pollution of our air and water resources, all of us—parents and children alike—need to participate in saving our planet. Only concerned and informed people can protect the Earth and turn things around. According to the famous African naturalist Baba Dioum:

In the end we will conserve only what we love, we will love only what we understand, and we will understand only what we are taught.

# Helpful Suggestions About Encouraging Your Young Scientist

You as a parent can have a tremendous influence on the way your child views science. Following are some suggestions and information about ways to encourage and support your young scientist.

## Be a participant, not a director.

Let your child select the experiments and activities that she wants to do. In many cases, her preferences will become apparent after a while. Does she constantly select outdoor activities? Does she enjoy collections? Is she especially interested in the stars and moon? Does she like experiments involving many manual steps? Let her do the experiments and activities on her own as much as possible while you act mainly as an assistant. Let her deviate from the given procedure or planned activity if she chooses, so that she can be free to explore. Science needs to be experienced, so it is especially important that children are given the opportunity to answer their own questions.

## Don't teach—let your child learn.

It is not important at this stage that your child understand all of the technical aspects of each experiment or activity; what is important is that he find science fun and thinking about science exciting. Encourage him to think about what is happening in his experiment or during his activity. Although an explanation is provided, it is much more important for him to think about it than to know the real reasons. The best way for children to learn science is with hands-on experiences, accompanied by conversations about what is happening.

## Plan the activity or experiment in advance.

Once you and your child have decided on a particular experiment or activity, make sure that you acquire all of the necessary materials in advance, improvising whenever necessary. Organization is a critical skill for any scientific endeavor; if you wait for the last minute, it is easy for your child to lose interest as you rummage through the cabinets looking for what you need. It is always a good idea to take along some piece of equipment or other needed items on field trips, such as inexpensive children's binoculars, magnifying glasses, a handheld microscope, a telescope, empty egg cartons for collecting rocks, or jars for collecting plant or water samples. Kids in general enjoy hands-on activities and like gadgets of all kinds. By using the different tools that are needed for science and nature studies, children will become at ease with the equipment and will also build manual skills.

## Use simple equipment and procedures for each experiment.

Although it may be tempting to obtain more advanced equipment and do more complicated experiments, at this stage of your child's development, more involved experiments may be counterproductive. The experiments and activities should be challenging but certainly within your child's ability to do and also understand. If she is unable to participate, she will likely become easily distracted and lose interest. Keep everything simple and straightforward so that her natural curiosity is engaged.

### Start a science/nature supplies box.

Collect and save recyclable items like plastic bottles and containers, cardboard, egg cartons, cereal boxes, shoe boxes, jars both small and large, cloth (from old clothes, for example), washed measuring caps from medicine containers, medicine droppers, etc. As you work your way through the different experiments and activities, you'll begin to look at these items in a different light.

### Record your experiments in a notebook used solely for that purpose.

Children often like to keep journals about what they do, so your child may enjoy describing in words or drawings his experiments or field trips. You can emphasize the importance to scientists of recording results of experiments as well as observations. All scientists use laboratory notebooks of some sort to keep their data. Having a notebook used solely for your scientific endeavors with your child also underscores the importance of this time that you spend together.

### Ask questions about what is happening and about what your child is observing along the way.

It is important to encourage your child not only to develop an interest in science and nature, but also to find that it is exciting to think about the experiments and activities. Try to ask a few questions along the way, such as "What do you see happening? Has there been a change in color? What does it feel like? Does it float? What are the properties of this solid? Does it dissolve? What happens when something dissolves?" Record your child's answers to these questions, showing her the importance of observing in detail the results of experiments. Do the same for field trips or visits to the museum. Try to pose questions that develop your child's observational skills as well as start her thinking about the topic. During a field trip to a forest, for example, questions like "I wonder why trees have leaves" may start your child thinking about what the leaves may be doing. The exact right answers are not important: At this point what is relevant is that you are helping your child to enjoy thinking about the world and the way it works.

## Take science and nature with you.

Use opportunities as they present themselves to continue the discussion of an experiment or activity that you have shared with your child. For example, if you have already observed water droplets form on the outside of a jar containing ice (Exercise #38), you can discuss the wetness of the outside of a glass of ice water at a restaurant, or the water droplets on the outside pane of a window during a hot day.

When your child expresses an interest, seize the opportunity to create activities and experiments that you can both share and that involve the interest. For example, if he becomes engrossed observing a butterfly or asks where they go during the winter, use libraries, bookstores, nature centers, zoos, and museums to learn together about butterflies. Plan a trip to a park, looking expressly for butterflies and observing their behavior. Learn about their life cycle, why they land on flowers, where they live, what they eat, who their predators are, etc. Build on his interests so that his fascination with the natural world will grow.

# Other Helpful Aids

There are many ways to encourage and develop your child's interest in science and nature other than performing experiments at home: Visit libraries and bookstores to look for information about special interests; find videos and television programs on these subjects; or, if you have access, surf the Internet—especially the World Wide Web—for relevant information. You can even join an organization or club that is involved in topics your child has found interesting. If she enjoys watching and studying birds, for example, you can join the local Audubon chapter and receive magazines and newsletters that you can share with your child.

**Reading** is an important part of learning science. Even if your child is not yet reading or just starting to, exploring libraries and bookstores or looking up topics in encyclopedias and dictionaries together can be fun and exciting for your child. It will help your child get into the habit of finding answers to questions by referring to appropriate written material. This will encourage her to explore and enjoy libraries, as well as acquaint her with science books. Much of the science anxiety that I have encountered in the classroom, especially in girls, related to having very little contact with science, as well as a misconception that it is too difficult to understand. By exposing children early on to science that they can understand and enjoy, you will not only help to alleviate fears but also encourage an interest in science and nature.

Today, more than ever before, people are learning through **pictures, diagrams, photos, videos,** and **CDs** in vibrant living colors. Children especially appear to be drawn by these other forms of communication. Search your library, bookstore, or video store for nature and environmental topics to learn about an area of interest to your child. *National Geographic,* for example, has made many videos for kids about different animals and fish as well as places such as forests and deserts. Many nature photographers have also composed books on numerous nature and environmental subjects.

The **Internet,** particularly the World Wide Web, is rich with information on the environment and many other scientific topics. Many of these Web sites are designed specifically for children. If you have access to the WWW, you will find a wealth of information in the form of text, pictures, photos, exhibits, and even videos.

I have included specific references and Web sites that are relevant to particular experiments and activities throughout this book. Listed here are a few general references and Web sites that you may also find interesting.

## References

The appropriate audience for each resource is indicated in parentheses: adult, children, both (children and adults), or teens.

*Helping Your Child Learn Science*
N. Paula
Office of Educational Research and Improvement, Washington, D.C., 1992 (ED330584)

*Parents and Schools: A Sourcebook*
A. L. Carrasquillo
Garland Publishing, New York, 1993

*Math, Science, and Your Daughter: What Can Parents Do?*
P. B. Campbell
U.S. Department of Education, Washington, D.C., 1992 (ED350172)

*The Kid's Nature Book* (both)
Susan Milord
Williamson Publishing, Charlotte, VT, 1989

*Hands-On Nature* (both)
Jenephur Lingelbach
Vermont Institute of Natural Science, Woodstock, VT, 1986

*Nature in a Nutshell* (both)
Jean Potter
John Wiley and Son, Inc., 1995

*Good Earth Art; Environmental Art for Kids* (both)
Mary Ann Kotz
Bright Ring Publishing, 1991

*Doing Children's Museums*
Joanne Cleaver
Williamson Publishing, Charlotte, VT, 1992

## Web Sites

National Parent Information Network (NPIN)
URL: http://ericps.ed.uiuc.edu/npin/npinhome.html

University of Minnesota at St. Paul (issues on education and health of children)
URL: http://www.fsci.umn.edu/cyfc/

Hands-On Science Centers Worldwide with Reviews
URL: http://www.cs.cmu.edu/afs/cs/usr/mwm/www/sci.html

## SECTION 1

# Shapes, Colors, and Light

People of all ages enjoy experiments that involve shapes, colors, and light. Children's fascination with shapes and colors is noticeable almost from the beginning—in mobiles, building blocks, ring towers, and more. And what child—or grown-up—doesn't look for rainbows from time to time? Extending these interests into the natural world, where shapes, colors, and light abound, will not only be great fun for your child, but will also be a valuable learning experience.

# Fun with Food Colors

Most children really enjoy playing with food colors. In the activities described here, you can use just a few common household items and food colors to introduce your child to some basic principles about liquids.

## 1. The Color Purple

**The Main Idea:** This exercise demonstrates how primary colors can be used to make many others.

**Things You'll Need:**

- 8 small white or clear plastic bowls or containers
- 4 food colors (red, yellow, blue, green)
- 1 cup and $\frac{1}{4}$ cup measuring cups
- 1 spoon for stirring

**How Long It Takes:** About 10 minutes

### What You'll Do

1. Add about 1 cup of water to 4 bowls.
2. Add 4 drops of each food color to a separate bowl and stir so that you have 4 different colored solutions of water. (If you use tablets, dissolve the tablets in bowls.)

3. Take an empty container and add $\frac{1}{4}$ cup of blue water to it. Add the same amount of yellow water and observe the color change.
4. Below are some other combinations that you might try. Invite your child to mix any colors she likes.

| *Mixing Colors* | *New Colors* |
|---|---|
| blue + yellow | green |
| yellow + red | ? |
| red + blue | ? |
| green + red | ? |
| colors of your choice | ? |

## Explanation

For older children especially, you can tell them about primary and secondary colors. There are three *primary* or basic colors: red, yellow, and blue. All other colors are called *secondary* colors and can be made by mixing the primary colors. You can use this color wheel to make secondary colors by combining the two colors next to the color you want. Also, if you mix colors that are opposite from each other on the wheel, your solution will turn gray.

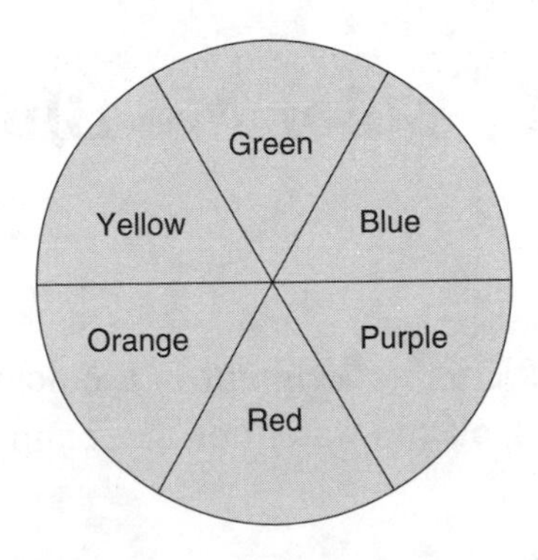


## Digging Deeper: Dark and Light Colors

Take one of your food colors and add four drops to one cup of water in one of your bowls. Do the same with another bowl so that you have two bowls containing the same color and the same concentration of food color. To one bowl add four more drops and compare the color. You can explain to your child that the color has darkened because of a greater *concentration*—that is, more food color. Now add water to the other bowl and note that the color gets lighter. The more *dilute*—in other words, the more water—the lighter the color.

## Underline your observation:

Add more food color and the color:
gets darker  gets lighter

Add more water and the color:
gets darker  gets lighter

# 2. Color Me Blue

**The Main Idea:** This experiment is a colorful way to show that some liquids like cooking oil and water don't mix and therefore form layers based on their densities.

**Things You'll Need:**

- Clear glass or jar
- Cooking oil
- Water
- Food colors
- Eyedropper

**How Long It Takes:** About 10 minutes

## What You'll Do

1. Pour water into your glass so that the water is about 1 inch high.
2. Slowly pour about the same amount of oil into the glass. Note that the oil floats on top of the water.
3. Use your eyedropper to gently release a few drops of blue food color into the oil.
4. Now gently push the blue droplets down into the water with the tip

of a spoon or fork handle. Watch them burst into a splash of blue as the food color spreads throughout the water layer.

5. Add another food color, such as yellow, and do the same as in Step 4. See the color of the water change to green. Remember that yellow and blue mix to form green.

## Explanation

Water and oil are two liquids that do not dissolve in each other. Liquids that do not mix are called *immiscible;* they form two layers with the lighter or less dense liquid (the oil) on top and the heavier or more dense liquid (the water) below. The blue droplets consist mostly of water and therefore will not disperse in the oil layer. Because water is heavier, the droplets will fall through the oil into the water layer where the blue color will spread out. You used a spoon to push the droplets down just to increase their rate of falling.

# 3. Layers of Colors

**The Main Idea:** This is another colorful way of demonstrating a scientific principle: When liquids do not mix, they separate into layers in the order of decreasing densities (lightest on the top and heaviest on the bottom).

**Things You'll Need:**

- Tall transparent glass or container
- Olive oil or other cooking oil
- Red or blue food color
- Water
- Glycerol
- Pancake syrup

**How Long It Takes:** A few minutes

## What You'll Do

1. Pour syrup about 1 inch high into a tall glass.
2. Pour about the same amount of glycerol followed by red or blue colored water and olive oil.
3. Identify the layer and its color for the diagram given below.

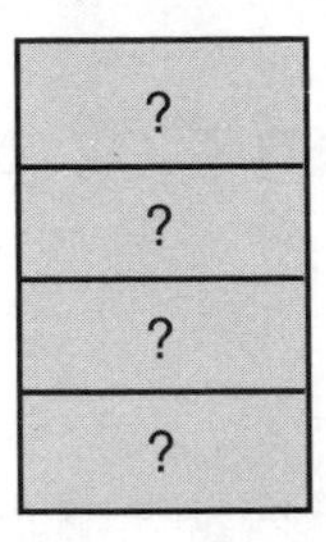


## Explanation

The liquids that you used do not mix. Therefore, they remain as layers according to their densities; the most dense liquid, the syrup, remains on the bottom, and the least dense, the olive oil, stays on top with glycerol and water in between. *Density* is a measure of the weight of a given amount (volume) of liquid, solid, or gas. The lighter liquids float on top of the heavier liquids.

# 4. Sink or Swim

**The Main Idea:** Here is a simple device you can make that tells you how the densities of different liquids compare.

**Things You'll Need:**

- 3 transparent glasses
- 3 straws
- Modeling clay
- Olive oil or other cooking oil
- Water
- Salt
- Red and blue food colors

**How Long It Takes:** About 15 minutes

## What You'll Do

1. Fill two glasses about $\frac{2}{3}$ of the way up with water. Fill the third glass about $\frac{2}{3}$ of the way up with cooking oil.
2. Add red food color to one glass of water and add blue color to the other. Then add about 1 tsp of table salt to the blue water. Stir until the salt is almost all dissolved.

3. Make a *hydrometer*, an instrument that compares densities of liquids with that of water, by attaching modeling clay to three straws as shown below.
4. Immerse your hydrometers in the liquids and mark on the straws the liquid level for each.

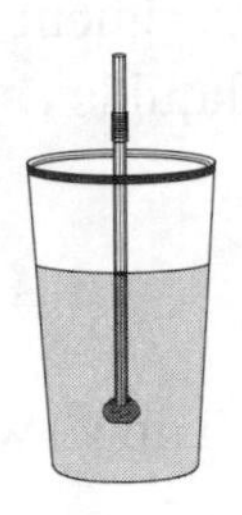

## Liquid Level

| | |
|---|---|
| Oil | ? |
| Red water | ? |
| Blue salt water | ? |

## Explanation

The hydrometer will sink the most for the least dense liquid, the cooking oil, and will float highest for the most dense, the salt water. That is why it is easier to swim in salt water (the ocean) than freshwater (a lake or river).

# 5. Rising Rainbows

**The Main Idea:** This experiment shows how water can separate the components of a mixture of liquids as the water rises through paper.

**Things You'll Need:**

- Coffee filters
- Flat container (like Tupperware®)
- Food colors
- Eyedropper
- Small cup (like the measuring cup on a medicine bottle)
- Tap water

**How Long It Takes:** About 1 minute

## What You'll Do

1. Mix 1 drop of each of 4 food colors—blue, green, yellow, and red—into the small cup. The resulting mixture should be gray-green after it is stirred or swirled.
2. Use your eyedropper to transfer 1 small drop of this mixture to about $\frac{1}{2}$ inch below the edge of a coffee filter.

3. Add enough water to the container so that its level is about $\frac{1}{4}$ inch high.
4. Gently place the coffee filter upside down so that the edges are immersed in water. Be careful not to wet the drop of food colors. Observe the water start immediately to climb up the filter and the colors separate before your eyes, creating a rainbow of colors.

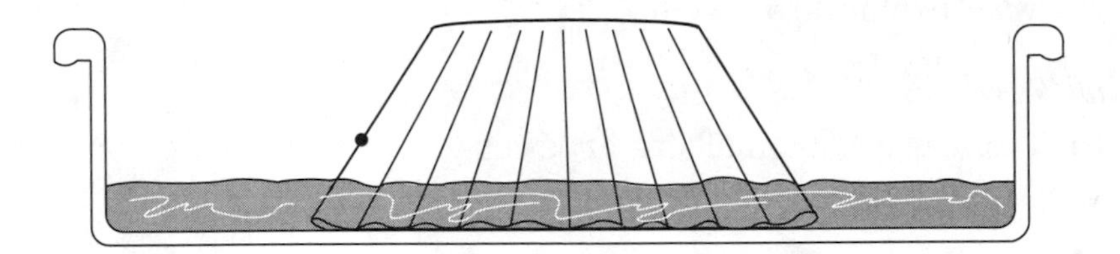

## Explanation

When the edge of a paper product like filter paper or a paper towel is immersed in water, the water will move upward. The dye that dissolves in water the fastest will travel with it, ahead of the other dyes. For older children, you can mention that this is a technique called *chromatography,* which scientists use to separate chemical substances.

## References

*175 More Science Experiments to Amuse and Amaze Your Friends* (both)
Terry Cash
Random House, New York, NY, 1991

*Solids and Liquids* (both)
David Glover
Kingfisher Books, New York, NY, 1993

*202 Science Investigations* (both)
Marjorie Frank
Incentive Publications, Nashville, TN, 1990

## Web Sites

*The Exploratorium* (children)
URL: www.exploratorium.edu/

*Kids Web*
URL: http://www.infomall.org/kidsweb/

# Shape, Rattle, and Roll

Shapes play a major role in nature, helping plants and animals adapt to their environment. For example, things float or sink, fly or fall, and are fast or slow depending on their shapes. Children also use shapes to make their own flying machines.

## 6. Ship Shape

**The Main Idea:** By changing the shape of an object, you can make it sink or swim.

**Things You'll Need:**

- Modeling clay or Play-Doh®
- A cup or bowl
- Tap water

**How Long It Takes:** About a minute

### What You'll Do

1. Form a ball out of clay or Play-Doh®.
2. Place it in a cup of water and watch the ball sink.

3. Remove the clay and reshape it so that it resembles a boat, like a canoe. Place it in the water again and watch the clay float.

## Explanation

An object floats only if the weight of the water displaced by the object is equal to the weight of the entire object. If the weight of the water moved aside by the object is less than the weight of the object, it will sink (Archimedes' principle). When the clay was in the shape of a ball, it was compact, so it displaced less water than when you spread it out into the shape of a boat.

## Digging Deeper: Gaining Weight

Use an eyedropper to put water in your clay boat. As it gets heavier, it will become more immersed in the water until the boat sinks. At that point it weighed more than the weight of the water it displaced. What happens when real ships become waterlogged?

You can also use a small capped jar that floats. Add a little water to the jar, close the cap, and place the jar back into the water; the jar sinks farther down as it displaces more water because of its weight gain. Add enough water until the jar sinks completely. At this point the jar weighs more than the weight of all the water that it moves aside.

**Test other items around the house to see what sinks and what floats. Can you make items that float sink, and those that sink float?**

# 7. Fly Away

**The Main Idea:** Shapes of objects not only determine their ability to fly, glide, or fall, but also determine their flight patterns.

**Things You'll Need:**

- $8\frac{1}{2}$-by-11-inch sheets of paper

**How Long It Takes:** Less than a minute

## What You'll Do

1. Crumple a sheet of paper in the form of a ball.
2. Give your child another sheet of paper, and you hold (or carry) the ball.
3. Both of you hold the paper at the same height and then release.

## Explanation

If we lived in a vacuum (no air), both sheets would reach the floor at the same time (a discovery made by Newton). Air, however, slows down objects—particularly those that are spread out.

# 8. Let's Twist Again

**The Main Idea:** The shapes of flying objects determine their flight patterns. Your child can easily make a paper rocket that will move smoothly through the air.

**Things You'll Need:**

- $8\frac{1}{2}$-by-11-inch paper

**How Long It Takes:** A few seconds

## What You'll Do

1. Roll an $8\frac{1}{2}$-by-11-inch piece of paper lengthwise into the shape of a cylinder.
2. Squeeze and twist one end into the shape of a nose.
3. Throw your "twister" with a little force, and it will glide for some distance. If you allow it to roll off your fingers, it will also rotate (twist) as it flies.

## Explanation

What is great about a twister is that it can be made by preschoolers without adult help. Because of its shape, air passes over it smoothly with less drag.

## Digging Deeper: Of Birds and Planes

There are many kinds of paper airplanes that you can make from simple sheets of paper. Make several of these and compare how they fly. The more spread out, the more they will glide, and the slower they will fly. The more streamlined, the less drag there is, and the faster the plane will travel. Note how different birds fly: Large hawks and eagles have wide wings and wing spans that they use to glide high in the sky on air currents, while birds like pheasants have short, broad wings for quick and steep takeoffs. Observe the flight patterns of different birds in your community. How does the shape of their wings relate to their flight patterns? If you like insects, compare their flight patterns and the shapes of their wings.

## References

*Physics Experiments for Children* (children)
Muriel Mandell
Dover Publications, Inc., New York, 1968

*Projects with Air* (children) and *Projects with Water* (children)
John Williams
Gareth Stevens Children's Books, 1990

*Audubon Adventures* (children)
Children's nature newsletter released by the National Audubon Society
700 Broadway, New York, NY 10003

*Hands-On Nature* (both)
Jenephur Lingelbach
Vermont Institute of Natural Science, Woodstock, VT, 1986

# You Light Up My Life

The way families flock to Fourth of July celebrations is an indication of our enjoyment of lights and colors. Following are some experiments that are fun to look at and, at the same time, reflect some important principles about light.

## 9. The Bright Colors of the Rainbow by ROY G. BIV

**The Main Idea:** In this experiment, you will see that white light is made up of a rainbow of colors. The first letters of the colors that make up visible light are ROY G. BIV: red, orange, yellow, green, blue, indigo, and violet.

**Things You'll Need:**

- Baking pan or tray
- Small handheld mirror

**How Long It Takes:** A few minutes

### What You'll Do

1. Fill the pan with water to just below the top.
2. Place the pan in bright sunlight not far from a wall.

3. Lean a small handheld mirror against an inside wall of the pan and orient it so that bands of colors appear on the wall.
4. Record or draw the different colors and the order in which they appear on the wall. The next time you see a rainbow, compare the colors and their ordering.

## Colors of my rainbow:

### Record of Colors:

#1______________________________________________

#2______________________________________________

#3______________________________________________

#4______________________________________________

#5______________________________________________

#6______________________________________________

#7______________________________________________

## Explanation

When the sun's white light penetrates the water in the pan, the light breaks up into its component colors. The mirror reflects these bands of colors, or *spectrum,* onto the wall so that you can view them. The exact same process occurs whenever a rainbow appears; there, water droplets in the atmosphere serve to separate the sun's rays into their component colors.

# 10. The Rainbow Connection

**The Main Idea:** This experiment shows that white light is broken up into its component colors whenever it passes through a clear material like water, glass, or plastic. Rainbows are pretty easy to make. In the "Digging Deeper" section you will see what happens when you pass white light through colored water.

**Things You'll Need:** Tap water

- White cardboard or posterboard
- Clear plastic or glass cup
- Flashlight
- Food colors
- A dark room
- Countertop or tabletop

**How Long It Takes:** A few minutes

## What You'll Do

1. Fill the clear plastic cup with water and place it on the edge of a counter or table.
2. Fold the white cardboard or posterboard in half and place it down like a tent a few inches behind the cup.

3. Turn on the flashlight and darken the room. Shine the flashlight on the cup from below the countertop or tabletop. What appears on the board?
4. Move the flashlight back and then forward. What happens to the spectrum of colors?
5. Record the colors and their sequence. Compare this rainbow to the next rainbow you see.

## Explanation

When white light goes through a clear material like water, plastic, or glass, the light is slowed down by this medium and breaks up into its component colors. Red is slowed down the least and violet the most. A rainbow appears in the sky when sunlight passes through water droplets in the sky.

## Digging Deeper: Passing White Light Through Colored Water

Add a few drops of food color to water in several cups. Do the same experiment but, this time, pass the white light into red colored water, then blue, yellow, green, and any others you make up. Try to guess the colors that will appear. The colored water has the effect of filtering out or absorbing some colors and transmitting the rest.

# 11. Why Is the Sky Blue?

**The Main Idea:** By passing light into water with milk droplets, you can see blue light being scattered toward you so that the water looks blue. Scattering of the sun's light rays also accounts for the blue color of the sky during the day and the reddish colors during the sunset.

**Things You'll Need:**

- Flashlight
- Clear plastic or glass cup
- Milk
- Tap water
- A dark room

**How Long It Takes:** About a minute

## What You'll Do

1. Fill the clear glass with water and then add a little milk. The water should now look somewhat cloudy.
2. Shine the flashlight through the glass so that the light beam is parallel to the table.
3. Darken the room and look down at the cup. What color do you see?

## Explanation

White light contains the colors of the rainbow. The undissolved particles in the milk and water mixture scattered the blue light from the light of the flashlight. The sky looks blue during the day and reddish during sunset because the molecules of air in our atmosphere scatter the sun's light. During the day blue light is scattered by the atmosphere toward the Earth. If air did not surround the Earth, the sky would look black except when you looked directly at the sun. At sunset the sun's rays must penetrate more of the atmosphere, so blue light is scattered away from the Earth, and more of the reddish colors come through.

# 12. Reflections of My Mind

**The Main Idea:** This exercise demonstrates that light travels in a straight path and is reflected when it hits a smooth, shiny surface like that of a mirror.

**Things You'll Need:**

- Flashlight
- Mirror
- A dark room

**How Long It Takes:** About a minute

## What You'll Do

1. Place a mirror on the floor and turn on your flashlight.
2. Darken the room and shine the flashlight on the mirror. Observe the reflection on the ceiling. Keeping the light on the mirror, first move closer to it and then farther away. What happens to the reflection on the ceiling?

## Explanation

Light travels in a straight line from the flashlight to the mirror. It then bounces off the smooth, shiny surface of the mirror and appears on the ceiling. As you make the light beam more slanted by moving away from the mirror, the reflected light also becomes more slanted and moves farther away on the ceiling.

**Try doing the same activity outside at night, perhaps near a campfire, and follow the light beam in the sky.**

## Digging Deeper: Multiple Reflections and a Periscope

Cut 2 holes the size of a quarter near the top of a carton and near the bottom on the opposite side. Tape 2 mirrors onto the box at a 45° angle to the box and parallel to each other. Make sure there are no other holes in the box. Look through the bottom hole. Can you see around corners and above counters?

# 13. Me and My Shadow

**The Main Idea:** This activity explains what shadows are and how the Earth's rotation on its axis causes both daytime and nighttime. Understanding shadows and nighttime may make them less frightening.

**Things You'll Need:**

- Metal can or toy whose shadow you would like to see
- Flashlight
- A dark room

**How Long It Takes:** Less than a minute

## What You'll Do

1. Place a metal can or toy on the floor in the middle of a room.
2. Turn the flashlight on and darken the room. Shine the light on the bottle or toy so that the flashlight is on the side of the object. Now slowly lift the flashlight so that the light is overhead. What happens to the shadow as the light shines from above?

## Explanation

Light travels in straight lines. When light hits an object like your toy or metal can, the light is bounced back. Thus, light is absent in the area behind the toy, and a shadow is cast. A shadow just means that there is no light. As you lift the flashlight, the shadow will get smaller simply because there is less area without light. When the beam is shining directly overhead, the object will not cast a shadow at all.

## Digging Deeper: Measuring Your Personal Shadow

With your child, select a sunny place in your backyard or in a park where she can stand with her back to the sun. Every few hours throughout the day, place a rock at the end of her shadow. Record where the sun is located each time you look for her shadow; just as with the can, her shadow gets smaller as the sun rises overhead. For older children, you might be able to demonstrate with 2 balls why the sun moves across the sky: Make a pen mark on the ball that represents the Earth and rotate it while you hold the other ball still, representing the sun. If the pen mark is the location of your child, she will be able to see why the sun appears in different positions of the sky.

## References

*Color and Light* (both)
John Williams
Gareth Stevens Publishing, Milwaukee, WI, 1992

*Color and Light* (both)
Smithsonian Institution
Gareth Stevens Publishing, Milwaukee, WI, 1993

*Light* (both)
Eiji Orii and Masako Orii
Gareth Stevens Publishing, Milwaukee, WI, 1989

*Catch a Sunbeam: A Book of Solar Study and Experiments* (both)
Florence Adams
Harcourt Brace Jovanovich, New York, 1978

*Sun Up, Sun Down* (children)
Gail Gibbons
Harcourt Brace Jovanovich, New York, 1983

*Bending Light* (children)
Pat Murphy et al
Little, Brown and Company, New York, 1993

## Web Sites

The Exploratorium (children)
URL: http://www.exploratorium.edu

Nye Labs (children)
URL: http:// www.seanet.com/vendors/billnye/nyelabs.html

Amateur Science from Bill Beaty (both)
URL: www.eskimo.com/~billb/

Kids Web (both)
URL: http://www.infomall.org/kidsweb/

## SECTION 2

# Kitchen Science

There are many experiments that you and your child can do with items commonly found in the kitchen. Your kitchen, replete with running water and a heat source, can become a wonderful laboratory where many scientific principles and facts can be discovered and appreciated.

# Icecapades

By doing some simple experiments with ice and water, you can enjoy learning some important properties of solids, liquids, and solutions and also understand some of the ways that ice behaves in our world.

## 14. Freezing Salt Water

**The Main Idea:** By using just water, table salt, and your freezer, you can see how salt stops water from freezing.

**Things You'll Need:**

- 2 ice trays
- Freezer
- Table salt
- Water
- Measuring cup

**How Long It Takes:** About 1 hour and 30 minutes, mostly spent waiting for the water to freeze

## What You'll Do

1. Dissolve 1 tsp of table salt in 3 cups of tap water.
2. Fill up one of the ice trays with the salt water and the other tray with just tap water. Use the same amount of water for each tray and use the same type of tray.
3. Place both trays in the freezer on the same shelf and check every 20 minutes or so to see when they are frozen. You and your child can each guess which ice will freeze first.
4. For an older child, you can also measure the temperature of the water (and ice-water mixture) each time you check on the trays.

## Explanation

Whenever solids dissolve in liquids or even when other liquids are mixed in, the freezing temperature of the original liquid lowers. Dissolving salt in water therefore lowers the freezing temperature of water below 32°Fahrenheit (or 0°Celsius). It will therefore take longer for the salt water to freeze than for the tap water because the salt water must get to a lower temperature in the freezer. If you guessed that the tap water will freeze first, then you were right. Applying your results, do you think that lake water will freeze before ocean water?

## Digging Deeper: Freezing Tap Water

Do the same experiment but use tap water and distilled water. Have your child taste the distilled water in order to recognize that it is different from tap water. Distilled water is pure, while tap water contains dissolved minerals from rocks and soil. In city and suburban regions, there are also low concentrations of pollutants, such as chloroform, from chlorination of the drinking water. Thus, the distilled water should freeze first.

# 15. Melting Ice with Salt

**The Main Idea:** In this experiment, you will see that salt can cause ice to melt. That's why we spread salt (calcium chloride) on ice during the winter. This experiment, like the preceding one, also demonstrates that salt lowers the freezing temperature of water.

**Things You'll Need:**

- 2 ice trays
- Tap water
- 1 tsp of table salt

**How Long It Takes:** Just a few minutes if you have ice in your freezer

## What You'll Do

1. Remove the 2 trays of ice from the freezer and generously sprinkle salt on top of the ice in one tray.
2. After just a few minutes, observe both trays; the salt will hasten the melting of the ice, and you will start to see water collecting. The tray without the salt will take much longer to melt.

## Explanation

A salt is a class of chemicals that are solids, and many of them dissolve in water. When salt dissolves, it causes water to remain a liquid. Thus, when you sprinkle salt on the surface of ice, the salt dissolves, causing the ice to melt.

# 16. Melting Ice with Pressure

**The Main Idea:** This quick experiment shows that when you press on ice, it melts. The activity explains why we are able to ice-skate.

**Things You'll Need:**

- Ice tray with ice cubes
- Freezer
- Gloves or towel

**How Long It Takes:** Just a few seconds if you have ice in your freezer

## What You'll Do

1. Remove the ice cubes from the tray and let them stand for a few minutes.
2. With gloves or a towel to protect your child's hands from the ice, ask her to press two cubes together for several seconds. When she stops pressing, the ice cubes should stick together.

## Explanation

When you press on ice, it melts, forming water. When you release the ice, the water freezes back to ice, connecting the two cubes. When we stand on ice, our body weight is pressing down, causing the ice under us to melt. We can then slide on top of the water—we can ice-skate!

**Add some food colors to the water in the tray before you freeze it so that you can have fun joining different colored ice cubes together.**

# 17. Ice Is Bigger Than Water

**The Main Idea:** This simple activity demonstrates a unique property of water: Water "expands," or takes up more space, when it freezes. Almost all other liquids do the exact opposite—that is, they shrink when they freeze. The activity demonstrates why ice floats.

**Things You'll Need:**

- Clear jar or plastic container
- Marking pen
- Tap water
- Freezer

**How Long It Takes:** About 1 hour for the water to freeze

## What You'll Do

1. Add water to a container without filling it up. Use your marking pen to show the level of the water.
2. Place your container of water in the freezer. When the water has turned to ice, remove the container and note that the ice is now above the original water level.

## Explanation

When water freezes, it forms a crystalline solid that has many empty spaces. That is why ice takes up more space than liquid water. This also accounts for why ice cubes float: Because ice takes up more space, its mass is more spread out, making ice less dense than water. Ice floats on lakes, rivers, streams, and oceans during frigid temperatures because it is less dense than water. What would happen to fish and sea mammals if water was the same as most other liquids, and became more dense when it freezes?

## Digging Deeper: "Raising Water Levels"

For older children, you can make this experiment more challenging. Take a 1-quart measuring vessel and fill it with water to the 2-cup mark. Pour out 1 cup into a jar or plastic container, which you will then place in the freezer. When the water has frozen, place the ice into the 1-quart measuring vessel that still contains 1 cup of water. Use a toothpick to push the ice down under the water, and note the level of the water in the measuring vessel. The level will be above the 2-cup mark because the 1 cup of water in the freezer expanded when it turned to ice.

# 18. Heat Wave

**The Main Idea:** In this experiment, you both will be able to see water in motion after it melts, and how it mixes with already existing water. You learn that the tiny particles of water we call molecules are constantly moving.

**Things You'll Need:**

- Ice tray
- Tap water
- Freezer
- Food colors
- Baking pan or a large Tupperware® container

**How Long It Takes:** About 1 hour and 30 minutes, mostly waiting for the ice to freeze and the colors to mix

## What You'll Do

1. Add water to your ice tray, leaving room for the expansion that occurs when ice freezes.
2. Add a few drops of food color to some of the cubes, skipping every other cube so that the colors do not mix. Choose two primary colors like yellow and blue for example. Place the ice tray in the freezer, making sure that it is level.

3. Cool some water by placing it in the refrigerator. Make sure that you use water to almost fill up your baking pan or Tupperware® container.
4. When the water in the ice tray has completely turned to ice, pour the cool water into your baking pan or container if it is not already in there.
5. Place one yellow ice cube at one end of the pan or container, and a blue colored cube at the opposite end. Observe the ice melting as the water around the cube becomes colored. As the water warms, it eventually turns green from the mixing of the blue and yellow.

## Explanation

As soon as the ice is taken out of the freezer and placed in the cool water, the ice starts to warm and melt. Because the ice is colored, you can actually see colored water forming. Because the water is cool in the container, the region around each ice cube will have the color of the cube. When the water warms, the blue and yellow colors mix to form green. The tiny particles or molecules that we cannot see are always moving, but they move very slowly when cool and more rapidly when warm. Thus, as the water warms, the molecules move faster and the colors mix.

## Digging Deeper: Speeding Things Up

Do the same experiment but, this time, place the cubes in a container with warm tap water. The melting and mixing will occur at a faster pace because the colored water will warm more rapidly and, therefore, the molecules will spread out faster. To see the effect of the water temperature, use two containers—one with cool water and the other with warm water—and perform the experiment on both containers simultaneously.

## References

*Solids and Liquids* (both)
David Glover
Kingfisher Books, New York, 1993

*Flying and Floating* (both)
David Glover
Kingfisher Books, New York, 1993

*Water* (both)
John Williams
Gareth Stevens Publishing, Milwaukee, WI, 1992

*Science with Water* (both)
Helen Edom
Usborne Publishing Ltd., Belgium, 1992

## Web Sites

*The Exploratorium* (describes many hands-on experiments for kids)
URL: http://www.exploratorium.edu/

# Acids and Bases

Acids and bases are all around us. From acid rain to Alka Seltzer®, we are constantly hearing about, using, and interacting with these two classes of chemicals. No doubt your child will hear early on about acids in particular, and that information is often erroneous. For example, many children think that acids in general are dangerous chemicals; this is true, of course, for very strong acids that you can purchase in hardware stores or chemical supply companies. However, most acids and bases that we come in contact with are weak and are prevalent in our diets: Citric acids, for example, are in many fruits and vegetables. Also, many common pharmaceuticals such as aspirin are weak acids. It is a good idea for your child to learn early on that acids and bases are everywhere, and that these naturally occurring chemicals are not dangerous but in fact are important to life.

Your child can think about acids as being liquids, consisting mostly of water, which taste sour. Bases also are liquids that consist mostly of water, taste bitter, and tend to feel slimy.

## 19. Is It an Acid or a Base?

**The Main Idea:** In this activity, your child can actually see which everyday product is an acid or a base. You and your child can prepare an indicator solution out of red cabbage and then use it to test different samples. Red cabbage gives a dark reddish purple solution, which will turn red when combined with an acid, and green when combined with a strong base. In this way, your child will enjoy looking at the colors and will also understand that we come in contact with acids and bases every day. Your child will also realize that they are not all dangerous, but most are very beneficial.

**Things You'll Need:**

- 4 or more small jars or glasses
- One large jar or glass
- Saucepan
- Knife and chopping surface
- Strainer
- Eyedropper
- Red cabbage
- Distilled water
- Substances to test, such as vinegar, orange juice, distilled water, rainwater, bicarbonate of soda (baking soda), and milk of magnesia

**How Long It Takes:** About 1 hour and 30 minutes

## What You'll Do

### Part 1: Making the Indicator from Red Cabbage

1. Chop up a red cabbage into fine pieces.
2. Bring about 1 pint (about 1-half liter) of distilled water in a saucepan to a boil. Add the cabbage pieces to the boiling water and then remove from the heat.

3. Let the cabbage cool and then pour it into a strainer, collecting the dark reddish purple liquid in the large jar. Discard the cabbage pieces into a compost pile. The cabbage juice will decompose if you leave it for about 2 days and will turn a redder color. For best results, you should use your indicator within a day of its preparation.

## Part 2: Testing for Acids and Bases

1. Pour a small amount of the indicator (cabbage solution) into a small jar so that there is at least 1 inch of the solution in the jar.
2. Use the eyedropper to transfer each item to be tested (vinegar, orange or lemon juice, distilled water, bicarbonate of soda, etc.). Observe the color change.

| *Substance Tested* | *Color of Indicator* | *Acid or Base* |
|---|---|---|
| Vinegar | ______________ | _____ |
| Orange juice | ______________ | _____ |
| Distilled water | ______________ | _____ |
| Rainwater | ______________ | _____ |
| Tap water | ______________ | _____ |
| Baking soda | ______________ | _____ |
| Milk of magnesia | ______________ | _____ |
| Any others you choose | ______________ | _____ |
| | ______________ | _____ |
| | ______________ | _____ |

## Explanation

Red cabbage, like elderberries or blackberries, forms a solution that changes color when an acid or base is added to it. Red cabbage forms a dark reddish purple solution that turns red when an acid like vinegar is added: The stronger the acid, the redder the color. When a base like milk of magnesia is added, the solution turns green; less basic substances like baking soda turn the cabbage solution blue, and neutral substances (not an acid or a base) have no effect on the color at all.

## Digging Deeper: Getting It Together

At the end of your testing, you might find that your child expresses a strong desire to mix all of the solutions together. Use a large bowl and encourage her to mix the indicator with vinegar. What color do you see? Then add some bicarbonate of soda. (Warning: Blue dots will appear before your eyes.) As it falls on the solution, what color appears? There will also be a lot of fizzing as the acid and base react to form carbon dioxide.

# 20. Testing the Waters

**The Main Idea:** Here is an opportunity for your child to learn that all water in the world is not the same. He can also learn about acid rain and the pollution that causes it. The procedure is the same as in the preceding experiment.

**Things You'll Need:**

- 4 or more small jars or glasses
- Large jar or glass
- Saucepan
- Knife and chopping surface
- Strainer
- Eyedropper
- Red cabbage
- Distilled water
- Water to test, such as distilled water, rainwater, pond water, lake water (especially from an acidified lake if possible); and vinegar

**How Long It Takes:** About 1 hour and 30 minutes

## What You'll Do

### Part 1: Making the Indicator from Red Cabbage

1. Chop up a red cabbage into fine pieces.
2. Bring about 1 pint (about 1-half liter) of distilled water in a saucepan to a boil. Add the cabbage pieces to the boiling water and then remove from the heat.
3. Let the cabbage cool and then pour it into a strainer, collecting the dark reddish purple liquid in the large jar. Discard the cabbage pieces into a compost pile. The cabbage juice will decompose if you leave it for about 2 days and will turn a redder color. For best results, you should use your indicator within a day of its preparation.

### Part 2: Testing for Acids and Bases

1. Collect samples of rainwater, tap water, distilled water, bottled drinking water, and water from ponds and lakes. If acid rain is a problem in your region (the Northeast especially), try to obtain a sample. You could also substitute vinegar, since acid rain in many areas is as acidic as vinegar.
2. Pour a small amount of the indicator (cabbage solution) into a small jar so that there is at least 1 inch of the solution in the jar.
3. Use the eyedropper to transfer each item to be tested. Does the color change?

| *Sample* | *Color of Indicator* | *Acid or Base* |
|---|---|---|
| Rainwater | ____________ | _____ |
| Tap water | ____________ | _____ |
| Bottled water | ____________ | _____ |
| Pond water | ____________ | _____ |
| Lake water | ____________ | _____ |

## Explanation

Distilled water, which is pure, will be neutral and therefore will not change the color of the indicator. The other samples will be acidic or basic, depending on what is dissolved in them, and therefore a color change should occur. Tap water is generally slightly basic and thus should turn the indicator blue. Rainwater is naturally slightly acidic because carbon dioxide from the air dissolves in it, and should turn the indicator red. If there is pollution in your area from cars or power plants, or polluted air is carried in by the wind currents, you may have acid rain. If so, ponds and lakes in your region may also be acidified. These acids will produce an even redder color similar to the result you get when vinegar is tested.

# 21. Acids and Bases Unite!

**The Main Idea:** By combining vinegar (an acid) with baking soda (a base), carbon dioxide is formed and can be used to blow up a balloon. This experiment demonstrates a chemical reaction: When acids and bases are mixed together, something else is formed.

**Things You'll Need:**

- Vinegar
- Baking soda (sodium bicarbonate)
- A balloon
- 1-pint plastic bottle
- Small funnel

**How Long It Takes:** A few minutes

## What You'll Do

1. Fill about $\frac{1}{4}$ of the plastic bottle with vinegar. If the vinegar was refrigerated, let it warm to room temperature.
2. Pour as much baking soda as you can into the balloon, using the small funnel.
3. Letting the balloon hang next to the bottle, secure the balloon onto the mouth of the bottle.

4. Once the balloon is attached to the bottle, lift it (the balloon) vertically so that the baking soda drops into the vinegar. What happens as soon as they mix?
5. Observe the balloon.

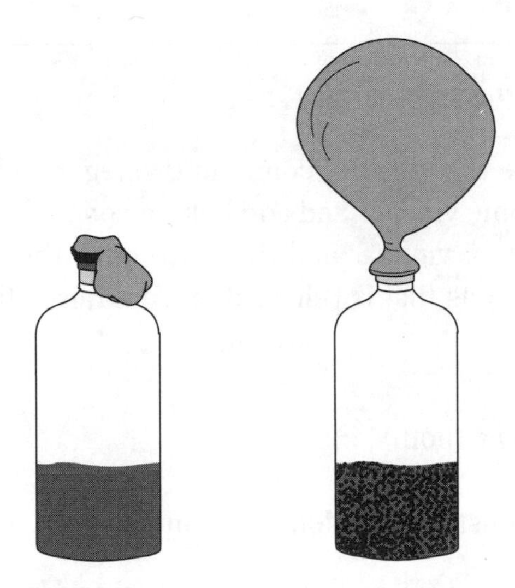

## Explanation

Acids and bases react to form other substances. When bicarbonate of soda (baking soda) and vinegar react, one of the products is carbon dioxide, a gas. The gas, just like air, can blow up a balloon.

The warmer the vinegar, the more quickly the balloon will expand. Fit as much baking soda into the balloon as possible, tapping the funnel if it becomes clogged.

## Digging Deeper: "Raisin" Raisins

A simple way to see the effect of combining vinegar and baking soda is to immerse a raisin in some vinegar and add baking soda. The raisin will drop to the bottom of the glass of vinegar and then rise once the baking soda is added. The carbon dioxide gas that is released is what raises the raisin.

## References

*How Science Works* (both)
Judith Hann
The Reader's Digest Association, Pleasantville, NY, 1991

*Hocus Pocus Stir and Cook: The Kitchen Science Magic Book* (both)
James Lewis
Meadowbrook Press, New York, 1991

## Web Sites

*Environmental Protection Agency* (provides information for kids on environmental issues like acid rain)
URL: http://www.epa.gov/

# Water, Water, Everywhere

Water really is everywhere; it constitutes about $\frac{2}{3}$ of our body weight and also occupies about $\frac{3}{4}$ of the Earth's surface. Water has many unique and special properties that all life depends on. The experiments described below will help your child understand and appreciate some of these properties.

## 22. I'll Race Ya!

**The Main Idea:** A very important property of water is that it dissolves many solids. In our bodies, water carries many substances to where they are needed. It carries nutrients up from soil into plants, enabling them to grow and produce food. This experiment will help your child understand what dissolving means and how solutions are formed.

**Things You'll Need:**

- 2 bowls or cups
- Sugar cubes
- 2 spoons for stirring (for preschoolers, use small spoons or stirrers)
- Stove top or microwave
- Tap water

**How Long It Takes:** About 30 minutes

## What You'll Do

1. Add about 1 cup of tap water to each of the 2 bowls.
2. In 1 bowl place a sugar cube and in the other, a crushed sugar cube.
3. Ask your child to pick one bowl, and each of you start stirring at about the same rate. Which sugar dissolved first? Encourage your child to taste a little of his solution.

How is the sugar solution different from the pure liquid water?

## Explanation

The dissolving solid (sugar) is called the *solute*, and the liquid (water) is the *solvent*. The *solution* is different from both the solute and solvent: In this case, it is not a solid and it tastes sweet.

When a solid dissolves, it breaks up into tiny particles that fit in between the water particles (molecules). Solutes dissolve faster when they are smaller. You can explain the reason to an older child: When you break the cube into tiny pieces, you increase the number of molecules of sugar that touch the water, making it easier for them to dissolve.

## Digging Deeper: Hot and Cold

Try repeating the experiment, but this time, use sugar cubes with cold water and warm water. How does the temperature of the water affect the dissolving process? Ask your child to think of different solids in the kitchen and see whether they will dissolve in water. For example, try salt, flour, or baking soda. Use a different liquid like cooking oil—will salt dissolve in it?

# 23. On the Road Again

**The Main Idea:** With a few simple items, you can watch water move through cloth fibers. This same property enables water to travel from the soil up into the roots and finally to the leaves of a plant, where the water is used to produce food.

**Things You'll Need:**

- 2 jars
- Old sock
- Tap water
- Food color
- Ruler (for older children)
- Scissors

**How Long It Takes:** A few hours, mostly spent waiting for the result

## What You'll Do

1. Use an old light-colored or white sock, maybe one that your child has outgrown. Cut out a long strip of cloth.
2. Place small dots of food color about every inch along the strip.
3. Fill a jar with tap water and immerse the end of the sock strip into the water. Hang the dry end over the edge of an empty jar that is lower than the jar containing water.

4. Make a sock chart by recording when the water reaches each dot as the water makes its way across the sock bridge into the empty jar. If, for example, you used green food color, after a few hours you will see drops of green colored water accumulate on the bottom of the empty jar.

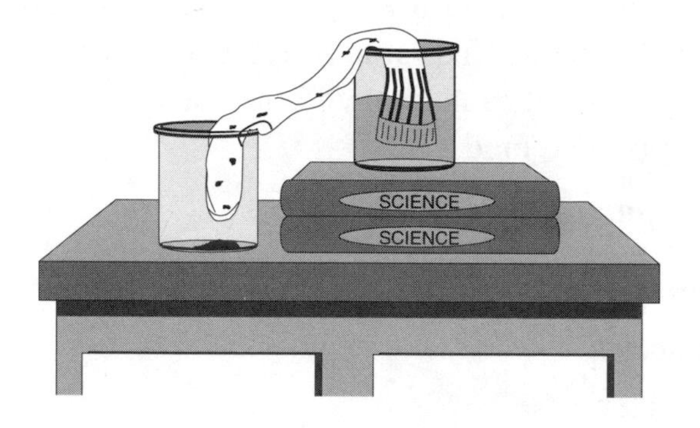


**Sock Chart**

| | *Spot 1* | *Spot 2* | *Spot 3* | *Spot 4* | *Spot 5* |
|---|---|---|---|---|---|
| Time Elapsed | | | | | |

## Explanation

Water is attracted to many solid surfaces more than it is attracted to itself. In this experiment, the water is attracted to the fibers in the sock and thus continues to move until it falls into the receiving jar. This is an example of *capillary action*.

## Digging Deeper: What Other Material Will Water Climb?

You can use paper towels or some other material to do the same experiment. Does water move through these other materials faster?

Try to help your child apply the principle learned here to some real-life situation. For example, you can ask your child this question: If you stand in a puddle of water without wearing boots, will the tops of your socks get wet?

# 24. Detergent Derby

**The Main Idea:** You can race boats with a drop of liquid detergent or soap that lowers the surface tension of water.

**Things You'll Need:**

- Cardboard or posterboard
- Scissors
- Pencil or pen
- Baking pan
- Tap water
- Liquid detergent or soap

**How Long It Takes:** A few minutes

## What You'll Do

1. Cut out a boat like the one shown below, or each of you can design your own.
2. Fill a baking pan full of water.
3. Place your boats gently on the surface of the water and place a drop of liquid detergent in the opening on the back of the boat. And the race begins!

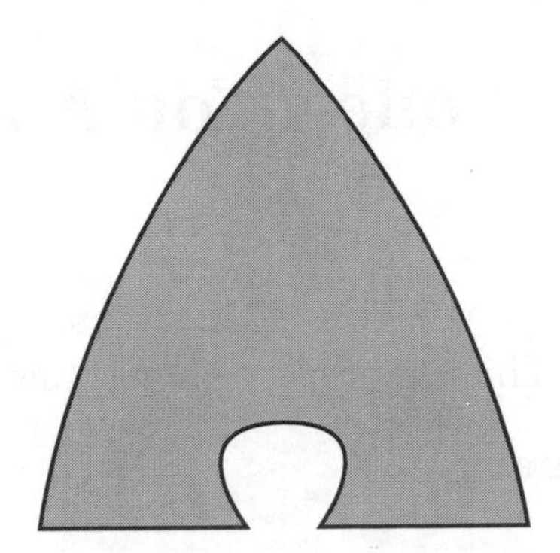

## Explanation

Water is very attracted to itself so that it forms a kind of "skin" on its surface called *surface tension.* Detergent breaks this skin, causing the boat to be drawn by the stronger surface tension in front of it.

**You can do the same experiment with just toothpicks or corks in a container with water.**

## Digging Deeper: Chasing the Blues Away

Add a drop of blue food color to a glass of water. Then add a few drops of liquid detergent and watch the blues run.

# 25. I'll Follow You Anywhere

**The Main Idea:** This experiment shows how a simple siphon works.

**Things You'll Need:**

- 2 jars
- Clothespin or clamp
- About a 2-foot length of rubber hose or tubing
- Faucet
- A few books

**How Long It Takes:** A few minutes

## What You'll Do

1. Place a few books on a counter near a faucet. Place a jar about 1 foot away from the books on the counter.
2. Fill the other jar almost to the top with tap water. Close one end of the hose with a clothespin or a clamp and fill the hose with water. Put your finger over the other end of the hose.
3. Place the end of the hose that you closed with your finger into the jar with water, being careful not to let any water spill out of the hose before you immerse it in the water. Place the other end in the empty jar and release the clamp. What happens as soon as the clamp is released? How can you use a siphon to empty an aquarium or a pool?

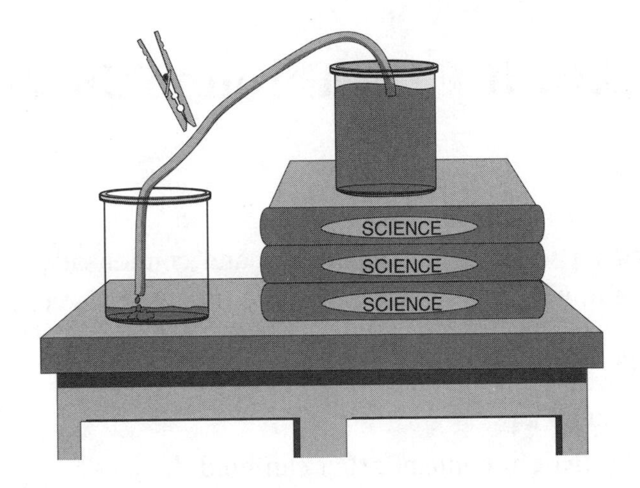


## Explanation

When you released the clamp, water flowed out because of its own weight (force of gravity), just as it flows when you tip a glass of water. Water continues to flow, however, because the air in the room (air pressure) keeps pressing on the water in the jar and pushing it up into the hose.

**Do the same experiment, but this time, do not use the books—keep the two jars level. Does the siphon work?**

# 26. It Flies Through the Air

**The Main Idea:** Both evaporation and condensation happen in this experiment, which helps us appreciate that water is constantly changing.

**Things You'll Need:**

- 2 small jars
- An airtight container that can hold the jars
- Salt
- Tap water
- A food color

**How Long It Takes:** Several days, mostly spent waiting to see the results

## What You'll Do

1. Fill one jar halfway with tap water and add a drop of food color.
2. Fill the other jar to the same level, dissolve as much table salt as possible, and then add one drop of the same food color.
3. Mark the level of the liquid in each jar. You can stick a white label on each jar and place a line at the water level. Make sure you also label the contents of each jar.
4. Place the jars in an airtight container or put them in a flat-bottomed bowl and cover it with a plastic wrap.

5. Observe the level of the water in each jar over a few days. Have the levels changed? Compare the darkness of the colors in the two jars. Are they the same?

## Explanation

Water will evaporate from the jar with plain tap water and condense into the jar with the salt water. The level of the salt water will be higher and also the color will be lighter because it is more dilute. Water always travels so that solutions become more dilute.

**Discuss with your child how water is always changing. It evaporates into the air from oceans, for example, and then comes down as rain and snow. Think of other examples around you, like the morning dewdrops on grass. What happens to them when the sun shines? Is it a good idea to water your garden during the heat of the day? It is important to apply to the world around us the principles that you demonstrate so that your child can see that science is everywhere—not just where you do experiments.**

## References:

*Science with Water* (both)
Helen Edom
Usborne Publishing Ltd., London, 1992

*Physics Experiments for Children* (both)
Muriel Mandell
Dover Publications, Inc., New York, 1968

*The Best of Wonderscience* (both)
American Institute of Physics
Delmar Publishers, an International Thomson Publishing Company, 1997

## Web Site

*Nye Labs* (Bill Nye, the Science Guy) (both)
URL: http://www.seanet.com/vendors/billnye/nyelabs.html

# The Toys

Young children love toys, and love to make toys themselves. Here you can enjoy making inexpensive toys with your child from common grocery items that she can play with and use for some time. In all of these activities, you can learn some interesting properties of the toys you create.

## 27. Rubber Ball

**The Main Idea:** By mixing two common household products, you can make a familiar rubber-like material with many unusual and entertaining properties.

**Things You'll Need:**

- Children's white glue that dissolves in water
- Borax
- Tap water
- Measuring spoons
- Small mixing bowl or container
- Small cup
- $\frac{1}{3}$ measuring cup
- Stirrer
- Waxed paper

**How Long It Will Take:** About 20 minutes

## What You'll Do

1. Add $\frac{1}{2}$ tsp of borax to a small cup and then add $\frac{1}{4}$ cup of tap water. Stir until the borax is completely dissolved.
2. To a small mixing bowl or container, add 1 tbsp of children's white glue and mix with 1 tbsp of tap water.
3. Add 1 tbsp of the borax solution to the glue mixture and stir for about 5 minutes with the stirrer. Does a new material form as you stir?
4. Place this rubber-like material on a piece of waxed paper for about another 5 minutes.
5. After the material dries, observe its properties. Is it anything like the glue or borax you used? Can you roll it into a ball? Does it bounce? Can you stretch it? What happens if you abruptly pull on opposite ends?

## Explanation

The rubbery material you made is an example of a *polymer.* The rubber and plastic in so many products that we use are polymers as well. Your polymer has very different properties from the original borax solution and white glue you combined: It bounces, stretches, and breaks.

**Before you mix the glue and borax solution, spend a little time with your child, noting the individual properties of these ingredients. What is their color and texture? Compare with the end product.**

# 28. Modeling Clay

**The Main Idea:** You can make modeling clay at home with just a few common ingredients. You can use the clay to model favorite animals and plants or to make imprints of leaves or animal tracks.

**Things You'll Need:**

- Flour
- Salt
- Vegetable oil
- Kneading board
- Mixing bowl
- Oven (optional)

**How Long It Will Take:** A few hours to complete, mostly waiting for the clay to harden

## What You'll Do

1. Mix $\frac{3}{4}$ cup of flour with $\frac{1}{4}$ cup of salt in a mixing bowl.
2. Add slowly $\frac{1}{4}$ cup of water to the bowl and then $\frac{1}{8}$ cup of vegetable oil, kneading to completely mix the ingredients.
3. Make any shapes you like with your clay. You can use a rolling pin to flatten the clay and then cut out squares that you can

or draw animal tracks on. Make holes on top of the squares so that they can be hung.

4. Let the clay dry for a few days or bake it in your oven at 250°F.

## Explanation

The proteins in flour form a substance called *gluten* that becomes hard when dry. Gluten not only makes good modeling clay but also bonds the ingredients of bread together.

**Kids get a special thrill out of making their own things that work. It gives them a sense of accomplishment to be able to make themselves what they see in stores.**

# 29. I'm Stuck On You

**The Main Idea:** Low on glue? You can make your own from a protein found in milk.

**Things You'll Need:**

- Vinegar
- Nonfat milk powder
- Tap water
- Baking soda
- 3 cups or jars
- Funnel
- Coffee filter or paper towels
- Teaspoon or stirrer

**How Long It Takes:** About 10 minutes

## What You'll Do

1. Prepare $\frac{1}{2}$ cup of nonfat milk from powder, following the directions on the package and using hot water. Pour into one of the jars or cups.
2. Stir 2 tbsp of vinegar into the milk and let stand for a few minutes until solids form.

3. Pour the liquid part out, keeping the solids.
4. Place the coffee filter or a paper towel in the funnel, and put the funnel in a second jar or cup. Pour the wet solids into the funnel so that some of the liquid can drain.
5. Remove the solids and place them in a third jar. To make your glue, stir in $\frac{1}{4}$ tsp of baking soda and $1\frac{1}{2}$ tsp of water.

## Explanation

The vinegar is used to remove the protein called *casein* from the milk (which is actually the glue).

# 30. Big Bubbles Are Blowin' in the Wind

**The Main Idea:** You can readily make a bubble solution that will allow you to blow bubbles a few feet in diameter. Adding glycerin to a soapy mixture is what makes the difference because it increases the surface tension.

**Things You'll Need:**

- Liquid soap or detergent
- Glycerin
- Distilled water
- Baking pan
- Stiff thin wire from a hardware store or hardware department

**How Long It Takes:** About 5 minutes

## What You'll Do

1. Use the wire to form an approximately 1-foot circle with a handle.
2. Mix together 1 part liquid soap or liquid detergent, 1 part glycerin, and 6 parts distilled water in a baking pan. Mix thoroughly.
3. Immerse the wire circle in the bubble solution and slowly remove the circle so that a thin film forms. Sweep the circle through the air and twist to free the bubble.

## Explanation

The glycerin adds *surface tension* to the solution. Surface tension is a force that resists the splitting of a liquid's surface. Thus, it adds strength to the bubbles. When you sweep the wire circle through the air, the tiny particles of air called *molecules* are pushing on the thin film, causing it to stretch and form a bubble.

**Because the bubble is so large, when it breaks, the soapy solution will splatter. It's best to make these bubbles outside and keep the solution from getting into your child's eyes.**

# 31. Good Vibrations

**The Main Idea:** With just some glass bottles and water, you can make a musical instrument and understand what causes high and low notes.

**Things You'll Need:**

- Several glass bottles
- Tap water
- A metal spoon

**How Long It Takes:** A few minutes or more if you hear a symphony.

## What You'll Do

1. Pour different amounts of water in each glass.
2. Strike the glasses and compare the pitch of the sound. Does the sound get higher or lower as you strike bottles holding more water?
3. Try to play a song that you both know. Try "Twinkle, Twinkle" (I'm sure you know the rest).

## Explanation

We hear sounds when air vibrates in our ears. The faster it vibrates, the higher the pitch of the sound. When the bottles are tapped with the spoon, water vibrates and then causes the air to vibrate. The bottle with the greatest amount of water will make a sound with the lowest pitch because the water vibrates the slowest. As the amount of water decreases, the pitch of the sound will rise.

You can also make music by blowing across the top of each bottle. In this case, the bottle with the most water will make a sound with the highest pitch; when you blow across the top of the bottle, you are making the air vibrate. The less air there is, the faster the vibration, and the higher the pitch of the sound. Which instrument—a flute or a tuba—will make a lower sound?

## References

*Mr. Wizard's Supermarket Science* (both)
Don Herbert
Random House, New York, 1980

*The Best of Wonderscience* (both)
American Institute of Physics
Delmar Publishers, New York, 1997

*Simple Science Experiments with Everyday Materials* (both)
Muriel Mandell
Sterling Publishing, New York, 1989

## SECTION 3

# Learning About Nature and the Environment

Young children in particular are naturalists; they investigate every berry, twig, rock, and insect they find. With these activities and experiments on nature and the environment, you can expand and support your child's fascination with the natural world, and also foster a respect and appreciation of it. You may even find that your youngster's enthusiasm awakens some of the sense of wonder you once felt, and you may view the world with a heightened awareness of its beauty and a desire to preserve it.

Since this subject area is so broad, I have broken down the activities in this section by their categories in order to provide you and your child with the widest possible variety of things to do.

# Tips for Studying Nature and the Environment

Here are some suggestions that are especially important in helping your child develop a concern and empathy for the living world.

## Be an observer, not a destroyer.

Try to leave a place you visit the way you found it. If you turn over a rock to inspect the organisms that live under it, return it to the way it was before you leave. Each area, no matter how large or small, is home to living creatures. Stay on paths in forests and on nature walks so as not to destroy the homes of wildlife.

## Take as little away as possible.

If you are collecting leaves or wildflowers, for example, take away as little as possible. Plants end up as part of the soil and then nourish new plants.

## Return animals and insects to their habitats.

If you catch ladybugs, fireflies, or a toad to observe and learn about, make sure you return them in good health to their homes.

## Intrude as little as possible on the lives of wildlife.

If you walk quietly and slowly, you will see so much more. Think of yourself as being a guest in someone else's home and act accordingly.

# 32. The Animals and the Beatles

Animals are especially important to young children; their books, shows, and toys all feature a variety of different animals. Children appear to be fascinated with the diversity of life: the amazing differences in the way animals look and behave. By studying animals with your child, you will help nurture an appreciation for the living world, and an understanding of each animal's importance in nature. You can also dispel myths about animals like wolves and bats, and learn how these animals play a vital role in maintaining the environment.

## #1. Following Animal Tracks

**The Main Idea:** You can learn much about the animals that cohabit in your community, or live in your favorite park or preserve, by observing their tracks. Go on a hike with your child to look for and follow animal tracks.

**Parents—Use common sense when following animal tracks. Be careful not to follow tracks that may have been left by a dangerous animal.**

**Things You'll Need:**

- Places where animals live
- Notebook
- Pencil

**How Long It Takes:** Optional

## What You'll Do

1. Look for animal tracks and follow them. Look especially around watering holes such as ponds, streams, or lakes. It also is a good idea to search when the ground is wet or after it snows so that the tracks are more visible and will show more details. Follow tracks to learn about what the animal was doing and thus gain some insight into its behavior.
2. Record where and when you see animal tracks, draw them, and then find out which animal made them.

| *Location of Tracks* | *Time Observed* | *Diagram* | *Animal* |
| --- | --- | --- | --- |
| ________ | ________ | ________ | ________ |
| ________ | ________ | ________ | ________ |
| ________ | ________ | ________ | ________ |
| ________ | ________ | ________ | ________ |

## References

*Crinckleroot's Book of Animals' Tracks and Wildlife Signs* (children)
Jim Arnosky
G. P. Putnam's Sons, New York, 1979

*Track Watching* (children)
David Webster
Franklin Watts, New York, 1972

*Snow Tracks* (children)
Jean Craighead George
Dutton, New York, 1958

## #2. A House Is Not a Home

**The Main Idea:** Loss of habitat is the primary reason for the extinction of animals worldwide. By studying the habitat of a familiar animal, a child can see that an animal's survival depends on the health of its habitat.

**Things You'll Need:**

- Notebook
- Writing supplies
- Modeling clay (optional)

**How Long It Takes:** Optional

## What You'll Do

1. Select a favorite animal or insect that lives in your community.
2. Identify where it might get its water, food, and shelter. How does it hide from its predators? Use the library or other sources of information like a zoo, nature center, or museum to help you.
3. Sketch your animal's habitat or use clay or other materials to make a model of it.

## Explanation

An animal's habitat is an area that contains food, water, and shelter. Some animals, like the cheetah, require large areas in order to survive, while others, like a rabbit, can survive in an empty lot. Animals that need large spaces tend especially to decline as human settlements continue to spread.

## Digging Deeper: Going, Going, Gone

Select an animal that is facing extinction, like the panda, or an animal whose population is declining, like the gorilla, orangutan, or elephant (Asian and African). Find out where the animal lives and learn together about its habitat. Figure out why it is disappearing. Think of some ways to stem the tide of extinction. Do parks and zoos help? Start a scrapbook that follows reports on the animal's survival and attempts to save it. Find out if there is an organization dedicated to its continued survival and join it.

## References

*Hands-On Nature* (both)
Jenepher Lingelbach
Vermont Institute of Natural Science, Woodstock, VT, 1986

*At Home in Its Habitat: Animal Neighborhoods* (children)
Phyllis S. Busch
The World Publishing Co., New York, 1970

*Farewell to Shady Glade* (children)
Bill Peet
Houghton Mifflin Co., Boston, 1966

*Animals in Danger* (children)
Janine Amos
Raintree Steck-Vaughn Publishing Co., Austin, TX, 1993

*The Endangered Species Handbook*, 1986 (adult)
Greta Nilsson
Write to: Animal Welfare Institute
P.O. Box 2650
Washington, D.C. 20007

*The Diversity of Life* (adult)
E. O. Wilson
Norton and Co., New York, 1992

*1001 Ideas for Science Projects on the Environment* (teens)
Marion A. Brisk
ARCO, New York, 1997

## Web Sites

*Animals* (both)
The San Diego Zoo
URL: http://www.sandiegozoo.org

*Animals* (adult)
World Resources Institute, Biodiversity Site
URL: http://www.wri.org/wri/biodiv/index.html

## #3. Learning About Insects

**The Main Idea:** Insects occupy all regions of the world. There are more of them than any other kind of animal, with over 900,000 known species, and they have survived for 400 million years. Children find them particularly interesting, perhaps because they are always around, and because there are so many different types. Their varied structures and behaviors have permitted them to successfully adapt to their environments, and thus be the most successful life-form on Earth. By observing, catching, and inspecting insects, children can see how animals adapt to their surroundings.

**Things You'll Need:**

- Hand lens
- Jars with covers
- Drawing materials
- Notebook

**How Long It Takes:** 1–2 hours, depending on your child's interest

## What You'll Do

1. Go into your backyard, a nearby park, or an open space to search for insects.
2. Listen for insect noises. Look for them under rocks, in flowers, and in the grass. Catch several in your clear jars for observation.
3. Using your hand lens, look for wings, shells, legs, eyes, antennae, and mouthparts. Draw the insects you see. How do the shapes of these parts help them survive? How does the shape of the mouth help the insect eat, or the length and shape of the legs help it move (such as those of a grasshopper, for example)?
4. Return the insects unharmed to where you found them.

## Explanation

Although insects are very diverse because of their individual adaptations, they all have 3 main parts: *head, thorax or middle section*, and *abdomen*. The head includes the eyes, antennae, and mouthparts that are particularly adapted to what the insect eats. The thorax contains legs (3 pairs), wings (usually 2 pairs), and muscles for movement. And the abdomen contains the main organs: the heart, holes for breathing, digestive system, and the reproductive organs. As you inspect the insects, try to determine with your child why the structures of the insects vary. For example, why does one insect have narrow wings while another has wide wings, or why is the mouth of a grasshopper different from that of a honeybee?

**Many insects can bite and sting. It is a good idea for you to become familiar with insects in your area before you engage in this activity with your child. Some books that will help you are listed below.**

## References

*Insects* (adult)
David Sharp
Dover Publications, New York, 1970

*Insects and Their Relatives* (adult)
Maurice Burton
Facts on File, New York, 1984

*How to Hide a Butterfly and Other Insects* (children)
Ruth Heller
Platt and Munk Publishing, New York, 1992

*Insects: A Close Up Look* (children)
Peter Seymour
Macmillan, New York, 1984

*Bugs in the System: Insects and Their Impact on Human Affairs* (adult)
May R. Berenbaum
Addison-Wesley, 1995

## Web Site

*Bees* (both)
Carl Hayden Bee Research Center's Global Entomology Agricultural Research Server
URL: http://gears.tucson.ars.ag.gov

## #4. The Life Cycle of Insects

**The Main Idea:** In this activity, you and your child can observe the life cycle of fruit flies. You can record and draw the changes that you see as the eggs eventually become mature adults in a process called metamorphosis.

**Things You'll Need:**

- Small glass jars
- Ripe fruit
- Paper funnel
- Small pieces of paper
- Cotton wool

**How Long It Takes:** A few weeks; a few minutes each day of observation and record keeping

## What You'll Do

1. Place a small piece of ripe fruit in a glass jar to attract the fruit flies. Then place a small piece of paper for the larvae to crawl on in the jar as well.
2. Take a piece of paper in the shape of a circle and fold it twice so that there are four quarters. Cut a small hole on the tip and then inset your finger into one side so that a funnel is formed. Place the funnel into the neck of the jar.
3. When 6 or 8 fruit flies have entered the jar, remove the funnel and loosely cover the jar with cotton wool.

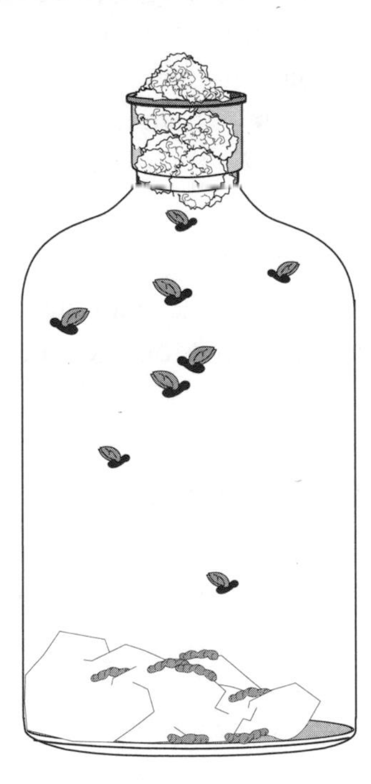

4. Look for eggs to be deposited. Watch them hatch and form larvae, which then pupate (form a cocoon) and give rise to the adult insects. Record and draw these changes.
5. Remove some of the young adult insects and place them in another glass jar to start a new generation.

## Explanation

After 6 or 8 fruit flies enter your jar, chances are you will have a mix of females and males; you can identify the females because they are bigger with a broader abdomen, while the male has a black-tipped abdomen. The deposited eggs hatch into larvae, which crawl over the pieces of paper in the jar. The larvae enter a resting state called the pupa. The adult insects will emerge from the pupae. Identify and draw these stages with your child. Most insect species—such as butterflies, bees, wasps, flies, and ants—go through the same metamorphosis.

## Digging Deeper:

Go on an insect hunt with your child, looking for insects at various stages of their lives. Look underneath leaves for eggs and under ledges for cocoons.

## References

*A Guide to Observing Insect Lives* (both)
Donald Stokes
Little, Brown, Boston, 1983

*The Strange Lives of Familiar Insects* (both)
Edwin Way Teal
Dodd, Mead and Co., New York, 1962

# 33. The Byrds

Bird-watching is a wonderful activity that you can enjoy with your child, regardless of where you live; birds coexist with humans in urban, suburban, and, of course, rural areas. Whether you have a simple bird feeder attached to your window or an elaborate backyard habitat, you can spend many pleasant hours observing the different bird species that visit during the year. Songbirds in particular are in serious decline worldwide, so they need all the friends they can get to help them survive! Your feeder, water station, nesting boxes, or backyard habitat will contribute to their continued survival. By observing birds, children can learn a great deal about how the anatomy and behavior of animals help them obtain food, water, and shelter, as well as how all living things play a role in the balance of nature. For example, insect-eating birds decrease the population of many disease-carrying insects.

## #1. I'm a Bird-Watcher

You and your child can spend many fun-filled hours just identifying the different bird species that come to your feeder or visit your neighborhood. Obtain a guide or handbook that describes birds that live in your area; there are many with clear photographs or drawings that will help you identify the birds you see. You can also contact a birding association that will suggest appropriate guides for your locality.

Record in a journal the kinds of birds that you see during the year, along with their characteristic features: What color(s) is it? How big is it? What is the shape of the bird's bill? What does it sound like? Does it eat on the ground, or will it eat from a feeder? What food does it prefer? Does it live here all year round, or does it migrate?

During the winter months especially, you can look for nests and determine what they are made of and which birds constructed them.

## #2. Making Bird Feeders Out of Recycled Materials

**The Main Idea:** There are many kinds of bird feeders made of different materials that you can buy at hardware stores, at garden supply centers, and from mail-order catalogs. You can also easily make your own feeders out of simple plastic soda bottles, milk cartons, or logs. Together, you and your child can take recycled materials and turn them into useful items that will help birds survive. Children tend to take a special interest in how well their custom-made bird feeder works over store-bought feeders. Encourage your child to think of other materials that could be used to construct feeders and try them.

**Things You'll Need:**

- 1-quart plastic bottle with cap
- Nontoxic paint
- Gravel or pebbles
- 2 chopsticks or dowels
- Scissors or sharp knife
- Cord

**How Long It Takes:** About 30 minutes

## What You'll Do

1. Use a scissors or sharp knife to cut the plastic bottle sidewards about 3 inches from the bottom. (Make sure your young child stays clear of any sharp instruments.) The slit should be about an inch long. Cut a similar slit on the opposite side of the bottle. Do the same about 3 inches from the top of the bottle, but on the two other sides.
2. Fill about 2 inches of the bottom of the bottle with gravel or pebbles.
3. Insert the chopsticks, dowels, or twigs from a tree through the slits. Make sure they are long enough for the birds to perch on as they are feeding.

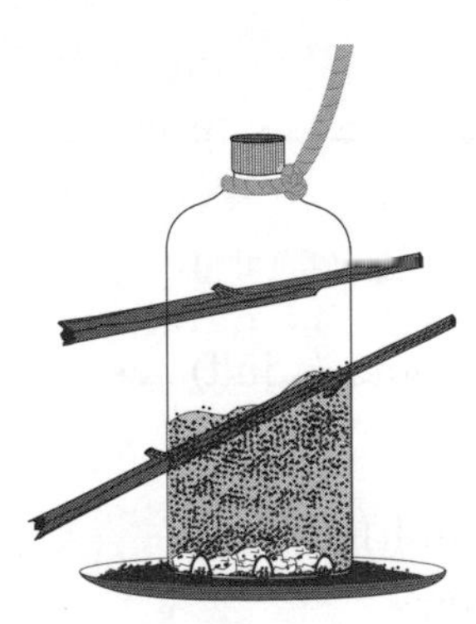

4. Press in the plastic above each slit so that the birds will be able to reach in and take the seeds.
5. Fill with bird seed and screw the cap on so that the seed does not get wet when it rains or snows.
6. Tie a cord around the neck of the bottle and hang it from a nearby branch or windowsill.

## Explanation

There are many feeders you can construct, including just an old baking pan placed on a porch. You can buy bird seed, prepare a variety of foods, or just put out leftovers. Feeding birds during the winter and nesting seasons is especially helpful when their energy output is high.

**You can determine the food preferred by each kind of bird that visits your feeder by filling the cups of an egg carton with a different seed or food and watching which one they select. Finches, for example, like thistle, while cardinals prefer sunflower seeds.**

## References


*Audubon Society Field Guides* (adult)
Knopf, New York

*Peterson Field Guides* (adult)
Houghton Mifflin, Boston

*The Birdhouse Book: Building Houses, Feeders, and Baths* (adult)
Don McNeil
Pacific Search Press, Seattle, WA, 1979

*National Audubon Society North American Birdfeeder Handbook* (adult)
Robert Burton
Dorling Kindersley, New York, 1992

*Feed the Birds* (adult)
Helen Witty and Dick Witty
Workman Publishing, New York, 1991

*The Tiny Patient* (children)
Judy Pederson
Alfred A. Knopf, New York, 1989

*The Nest: An Ecology Storybook* (children)
Chris Baines
Crocodile Books, New York, 1990

*Anna's Rain* (children)
Fred Burstein
Orchard Books, New York, 1990

*Handbook of Nature Study* (adult)
Anna Comstock
Cornell University Press, Ithaca, NY, 1911

## Web Sites

*National Audubon Society* (both)
URL: http://www.audubon.org/audubon/contents.html

*Cornell Laboratory of Ornithology* (both)
URL: http://www.ornith.cornell.edu/Feedback.html

*Birdlinks* (both)
URL: http://www.phys.rug.nl/mk/people/wpv/birdlink.html

*Birding on the Web* (both)
URL: http://www.birder.com

## Organizations

The American Birding Association
P.O. Box 6599
Colorado Springs, CO 80934

Audubon Society
950 Third Avenue
New York, NY 10022

You can also consult botanical gardens, nature centers, and local Audubon Chapters for information about birds in your community.

# 34. Environmental Science

Understanding and appreciating our environment are really crucial for this generation. Today's children will be living in a world where major changes will need to be made so that clean water and air, fertile soil, and energy will be available. It is important therefore that they view the environment differently from past generations: that they see the Earth's resources as not being inexhaustible and there for us to use up, but as limited and in need of protection. In these projects, your child can learn about some environmental problems and what he or she can do to help.

## #1. What is Biodegradable?

**The Main Idea:** In this activity, your child can see firsthand how some things we discard will naturally decompose in the ground, while others, such as plastics and glass, will not. This experiment also emphasizes the importance of recycling so that we are all not knee-deep in garbage.

**Things You'll Need:**

- Plastic bottle
- Metal can
- Newspaper
- Kitchen scraps (leave out meat products)

- Small area where these materials can be buried
- Shovel

**How Long It Takes:** About 30 minutes to bury materials with your child, and a few months or longer to observe results

## What You'll Do

1. Dig 4 holes about 2 feet apart and about 1 foot deep.
2. Bury the plastic bottle, metal can, newspaper (make sure it's not today's), and the kitchen scraps each in separate holes.
3. Uncover the holes after several weeks to observe the condition of the materials. How do they look? Which waste material has decreased in size? Continue monitoring the progress of the decomposition and record your child's observations.

## Explanation

There are millions of living organisms in the soil you used in your experiment; they include worms, insects, bacteria, and fungi. These organisms all participate in the breakdown of *biodegradable* materials. Matter from plants and animal is very biodegradable, followed by paper, metal cans, and most plastics (which are not considered biodegradable). The kitchen wastes will decompose first, returning nutrients to the soil. You can explain to your youngster how the plants grew by taking nutrients from the soil and that now some of these nutrients are being returned, allowing other plants to grow. The paper will start shredding and decomposing quickly, while it will take many years for the metal can to decompose. This hands-on learning experience will demonstrate the importance of recycling to reduce garbage.

Try to use soil that is rich in humus: The *humus* or black part of the soil contains decaying plant material that attracts decomposing organisms. Also, make sure the plot of land you use has not been sprayed with herbicides or pesticides.

## References


*The Great Trash Bash* (children)
Loreen Leedy
Holiday House, New York, 1991

*Recycle! A Handbook for Kids* (children)
Gail Gibbons
Little, Brown, and Company, 1992

*The McGraw-Hill Recycling Handbook* (adult)
Herbert F. Lund
McGraw-Hill Inc., New York, 1993


## Web Sites


*The EcoWeb*, University of Virginia
URL: http://ecosys.dr.dr.virginia.edu/EcoWeb.html

*Environmental Defense Fund*
URL: http://www.sun-angel.com.edf/edf.html

## #2. Water Erosion

**The Main Idea:** A major environmental problem worldwide is the loss of agricultural soil through water erosion. Erosion is the removal of topsoil by moving water or by wind. This process occurs primarily when trees and plants are cleared and the soil is left bare. In this activity, you and your child can observe the importance of plants in absorbing water, and also in anchoring the soil so that it is not washed away.

**Things You'll Need:**

- 3 trays
- Screens
- Soil
- Grass sod or other plants
- Watering cans
- Pails or large containers

**How Long It Takes:** 1–2 hours

## What You'll Do

1. Construct or obtain 3 trays; you can use thin plywood along with a screen attached to one side. Use putty or glue to plug spaces in the corners.
2. Fill 2 trays with fairly loose soil, and the third with soil and grass sod, or soil and other plants.

3. Use a box or thick boards to position the 3 trays on an incline. Position one of the soil trays at a greater angle than the other two (see diagram below).
4. Place receptacles below the trays to catch the running water.
5. Use a watering can to pour water over the trays. Observe the water dripping into the containers as well as the surface of the trays. Which tray is losing the most soil? Which one is losing the least? If it rains outside, will more soil be taken away from a hillside than from flat ground? If there are plants and trees on the hillside, will less soil be lost?

## Explanation

The tray with the uncovered soil and on the greatest incline will suffer the most erosion, the same way that hillsides that are cleared of their forests lose their soil very rapidly. Plants slow down erosion and therefore are important in keeping topsoil. Should farmers and gardeners leave their fields bare in the winter, or should they cover them with plants?

## Digging Deeper: Runoff and Rivers

Sprinkle the same amount of water on each of the 2 trays containing just soil and compare the amount and color of the water in the 3 containers. More water and soil will most likely come from the inclined tray with the uncovered soil, and the least water from the tray containing plants. This result is analogous to the flooding that sometimes occurs when there is very heavy rainfall in areas where many of the trees and plants have been removed to make room for cities, suburbs, and farms; or when large tracts of land are clear-cut by timber companies. What happens to the rain when it hits the ground? Some rain is absorbed by the soil, but in uncovered areas much of it flows along the surface and is called *runoff*. The runoff drags soil with it, eventually ending up in a river, stream, lake, or bay. This process explains why streams and rivers have a muddy color.

## Reference

*Saving Our Soil* (adult)
James Glanz
Johnson Books, Boulder, CO, 1995

## Web Sites

*The United States Department of Agriculture*
URL: http://www.usda.gov/usda.htm

*Soil, Water, and Climate*
URL: http://www.soils.umn.edu/

## #3. Composting

**The Main Idea:** Getting your child involved in composting is a great way for her to see and appreciate how decaying plant and animal matter fertilizes the soil so that other plants can grow. Composting also makes use of "garbage" and thus decreases landfill requirements.

**Things You'll Need:**

- Compost bin
- Shovel
- Topsoil
- Seeds for quick-growing plants, such as beans
- Flowerpots

**How Long It Takes:** About 6 weeks for compost to decay sufficiently and a few weeks to observe the difference between plants potted in compost containing soil and plants that are not

## What You'll Do

1. Build or buy a compost bin. You can use old lumber to build a bin, making sure there are spaces for air by leaving gaps between boards, or you can wrap chicken wire around 4 stakes firmly placed in the ground. Without oxygen, compost will acquire an odor reminiscent of a swamp.

2. Add mainly plant matter—for example, kitchen scraps (fruit and vegetable remains, coffee grounds, and egg shells), dry leaves, dry grass clippings, straw, dry woody plants, and sawdust. Avoid meat and cheese, which may attract large animals and also give your neighbors reason to complain.

3. Layer with dirt to help increase the rate of decay, and layer occasionally with twigs to bring in air.

4. Compost decays fastest when it is moist. Add water during dry spells and cover when rainfall is excessive.

5. Every few weeks, turn over the compost. Compost piles heat up as a result of the decomposition of the plant matter. Feel the interior of the pile. With older children, you can use a thermometer to measure its temperature.

6. After a few months, remove some of the decayed compost and mix it with dirt from your yard. Use this rich mixture to grow fast-growing seeds, such as green beans, in flowerpots. Do the same with dirt without compost added. In which soil did the seeds sprout first? For older children especially, observe the plant growth, recording every few days the length and appearance of their stems, branches, and leaves.

## Explanation

When plants decay, they return nitrogen and other nutrients to the soil. Compost is therefore a natural fertilizer that, when added to soil, will help plants grow. This activity also shows how garbage can be turned into a useful product.

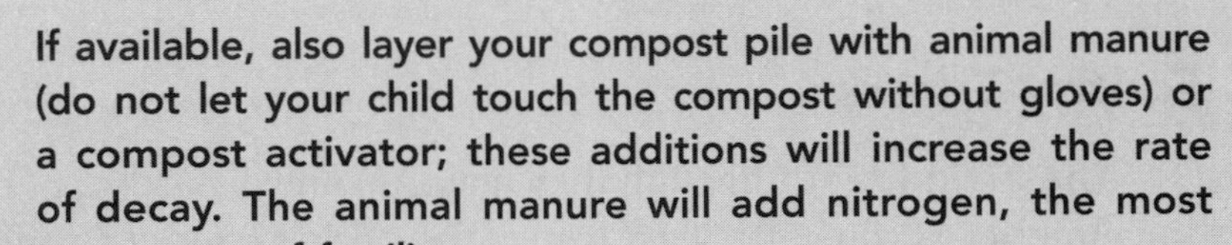

**If available, also layer your compost pile with animal manure (do not let your child touch the compost without gloves) or a compost activator; these additions will increase the rate of decay. The animal manure will add nitrogen, the most important component of fertilizer.**

## References


*Compost Critters* (children)
Bianca Lavies
Dutton, New York, 1993

*Let It Rot! The Gardener's Guide to Composting* (adult)
Stu Campbell
Garden Way Publishing, 1975

*Backyard Composting: Your Complete Guide to Recycling Yard Clippings* (adult)
Harmonious Press, P.O. Box 1865-100 Ojar, CA 93024

*The Compost Heap* (children)
Harlow Rockwell
Doubleday, New York, 1974

## Web Sites

*Alternative Agriculture News* (adult)
URL: http://envirolink.org/pubs/Alternative_Ag_News

*Farmer to Farmer* (adult)
URL: http://www.organic.com/Non.profits/F2F/

## #4. Energy Eaters

**The Main Idea:** Although they make up only 5 percent of the population, Americans consume 30 percent of the world's energy. Burning oil and coal causes acid rain and global warming. Here you and your child can determine how to reduce your use of energy and thereby tread more lightly on the Earth.

**Things You'll Need:**

- Small notepad
- Pencils

**How Long It Takes:** Track your energy use for one week

## What You'll Do

1. Older children can carry with them a small notepad and either jot down each time they use energy or write a summary after every few hours. You can do the same for 1 week.
2. At the end of the week, look at your patterns of energy consumption and brainstorm with your child on how your family can reduce its

use of energy. Your local electric company most likely has information that will be helpful in reducing your electric bill.

## Explanation

It is important for children to understand that using energy from fossil fuels is damaging to the environment and that all of us can help by using less energy. Find ways that your family can contribute to preserving the Earth, such as shutting off lights, carpooling, using mass transportation, insulating your home, and using more energy-efficient light bulbs and appliances.

## Digging Deeper: Water Worries

You can do the same exercise for your water use as well. Water resources are dwindling in many regions of the nation (especially the Southwest) as industries, farms, and cities all use more and more water. As the water levels of rivers and lakes decrease, the fish and wildlife suffer. Discover some ways that your family can use less water, such as not letting water run, taking shorter showers, having plants that use less water, and filling washing machines with clothes before you use them.

## References

*A Consumer Guide to Home Energy Savings* (adult)
Write to: American Council for an Energy Efficient Economy (ACEEE)
1001 Connecticut Ave., N.W., Suite 801
Washington, D.C. 20036

*99 Ways to a Simple Lifestyle* (adult)
Center for Science in the Public Interest
Doubleday, New York, 1977

*The Energy Saver's Handbook: For Town and City People* (adult)
Massachusetts Audubon Society
Rodale Press, Emmaus, PA, 1982

*Earth Child* (both)
Kathryn Sheehan and Mary Waidner
Council Oaks Books, Tulsa, OK, 1994

*1001 Ideas for Science Projects on the Environment* (teens)
Marion A. Brisk
ARCO, New York, 1997

## Web Sites

*The Energy Outlet* (adult)
URL: http://energyoutlet.com/

*Energylinks* (adult)
URL: http://www.epri.com/energylinks.html

*U.S. Department of Environmental Protection (EPA)* (both)
URL: http://www.epa.gov/

*Energy Ideas Clearinghouse* (adult)
URL: http://www.energy.wsu.edu/ep/eic/

*U.S. Geological Survey Water Use Homepage* (adult)
URL: http://h20.usgs.gov/public/watuse/wuqa/home.html

# 35. Plants, Soil, and Rock

Seeing a plant grow from a planted seed is especially exciting for youngsters. Activities and experiments involving plants and soil can help your child understand the interconnectedness of the living world and how important soil and plants are to us.

## #1. Plants Exhale

**The Main Idea:** This experiment shows how plants—just like people—exhale water during respiration. Ask your child to breathe on a mirror and then touch the water vapor that condensed on the surface. Then do the experiment, showing how plants also release water vapor from respiration.

**Things You'll Need:**

- Small potted plant
- Small pot with soil
- Aluminum foil or cardboard
- 2 glass jars

**How Long It Takes:** About 15 minutes to prepare; a few hours to observe

## What You'll Do

1. Obtain or grow 1 small potted plant. Place soil in another planting pot of the same size.
2. Water the soil in both pots and then use aluminum foil or cardboard to fully cover the soil.
3. Invert jars over the pots, as shown below, and place them side by side in the sun for a few hours. Check them periodically. Does water appear on the inside of one jar and not the other? Touch the inside of the jars to feel for wetness.

## Explanation

Plants, like animals and humans, inhale oxygen from the air and exhale carbon dioxide and water through their leaves. They respire through small openings in their leaves called *stomata*. The reason why you need a pot without the plant is to show that the water is not coming from the pot itself or from the soil; the plantless pot acts as a *control*. The only difference between these two pots is the presence of the plant, so the water must be coming from the plant. Try to explore this concept of a control with your child and why it is necessary in an experiment.

## Digging Deeper: Plant Respiration

Place a plant in a small jar with water; then set the jar down in a pan with water. Invert a larger jar over the plant and its container. Keep the jar away from light and observe the jar the next morning. Because plants inhale oxygen from the air, the water level in the inverted jar will rise to take up the room left by the absorbed oxygen.

## Reference

*Plant Experiments* (children)
Vera R. Webster
Children's Press, Chicago, 1982

## #2. Plant Stems and Food Color

**The Main Idea:** Children can see how plants bring water and nutrients to their branches and leaves with this experiment.

**Things You'll Need:**

- Celery stalks with leaves
- Jars or glasses
- Red and blue food colors
- Knife

**How Long It Takes:** About 10 minutes to set up and a few hours to let stand

## What You'll Do

1. Cut about an inch off the end of the celery stalks and let them stand in cold water for about an hour. Don't cut the leaves.
2. Put a few drops of food coloring or a tablet in a jar (try red in one and blue in another). Then fill the jar with about a $\frac{1}{2}$ inch of water.
3. Place a celery stalk in each jar or glass and let stand for a few hours.
4. Look for colors to show on the celery leaves and then cut the celery stems into 2- or 3-inch length pieces so that you can see where the colored water moved up the stalk. Carefully break the pieces apart and pull some of the colored tubes out so that you can observe them.

## Explanation

The tubes in the stems of plants carry water and dissolved nutrients from the roots up to the leaves, flowers, and fruits. In this experiment, the colored water in the jar travels up into the leaves.

## Digging Deeper: Looking for Tubes in Tree Branches

Find some twigs from different trees and place the twigs in colored water. After a few hours, cut them into small lengths and look for where the water has traveled up the stem.

## References

*More Than Just a Vegetable Garden* (children)
Dwight Kuhn
Silver Burdett Press, Morristown, NJ, 1990

*Hands-On-Nature* (both)
Jenepher Lingelbach
Vermont Institute of Natural Science, Woodstock, VT, 1986

*Earth Child* (both)
Kathryn Sheehan and Mary Waidner
Council Oak Books, Tulsa, OK, 1994

## #3. Planting Seeds

**The Main Idea:** Planting seeds and watching them germinate is especially exciting for children. With these activities, children learn where vegetables and fruits come from and how important fertile soil is to plant growth.

**Things You'll Need:**

- Seeds for fast-growing plants, such as beans
- Clear glass or plastic jars or containers
- Potting soil

**How Long It Takes:** About 10 minutes to set up and a few weeks to observe

## What You'll Do

1. Fill several clear containers with soil and plant the seeds near the edge of the containers.
2. Follow directions on seed envelopes as to where the containers should be placed for optimum growth and how they should be watered.
3. Observe the growth of the roots through the clear container. Notice how the main roots branch off into smaller and smaller roots.
4. You can transplant outside in larger pots, or in the ground if the weather is warm enough in your area. Continue to observe the plant growth.
5. For older children, record when the seeds germinated and the lengths of stems, leaves, and roots every few days. For preschoolers, you can sketch together the progress of your plant.

## Explanation

You will be able to see the growth of your seed above and below the soil in your clear container. The roots will develop into many branches, the smallest of which are *root hairs* through which water and nutrients are absorbed. Some roots, such as carrots, beets, and radishes, are edible.

## Digging Deeper: Inside a Seed

Soak seeds of bean, pumpkin, corn, and sunflower for a few hours. (Lima beans are especially easy to use.) Remove the outside layer of the seeds and carefully cut the seeds open. Show your child the small plants (embryo) in the seeds and explain that the rest of the seed is stored food that the young plant uses as it grows.

**Plant a garden with your child so she can see how different plants grow and produce fruits and vegetables. Closely observe the flowers that form and the small vegetables and fruits that appear.**

## References

*Growing Things* (children)
Angela Wilkes
Usborne Publishing, Ltd., 1984

*The Victory Garden Kid's Book* (children)
Marjorie Waters
Houghton Mifflin, Boston, 1988

*Let's Grow: 72 Gardening Adventures with Children* (both)
Linda Tilgner
Storey Communications, Inc., Pownal, VT, 1988

*Now I Know All About Seeds* (children)
Susan Kuchalla
Troll Associates, Mahwah, NJ, 1982

*Look, Mom It's Growing: Kids Can Have Green Thumbs Too*
Ed Fink
Countripede Books, Barrington, IL, 1976

## #4. Soil and Plant Growth

**The Main Idea:** This activity demonstrates how important the soil is to plant growth. People often view soil as just dirt, and have little understanding of how this thin outer layer of the Earth is necessary for our survival. In this activity, you and your child can grow plants, using topsoil and subsoil that you remove by digging down about 1 foot into the ground.

**Things You'll Need:**

- Large and small garden shovels
- Clear glass or plastic containers or jars
- Soil
- Fast-growing plants, like corn or beans

**How Long It Takes:** About 10 minutes to set up and a few weeks to observe

## What You'll Do

1. Find rich topsoil from a garden or use potting soil. **Do not** use soil that has been treated with pesticides or herbicides or has been fertilized with synthetic fertilizers—these chemicals can be dangerous to your child! Fill several containers with the fertile soil.
2. Dig a hole about 1 foot deep or until the soil clearly becomes lighter in color and changes texture. Fill several containers with this subsoil.
3. Plant corn or beans or some other type of fast-growing seeds in these containers. Care for them according to the directions on the seed packets.
4. Observe how the different plants germinate and grow: Which ones germinated first? How do their roots compare? How much does each plant grow in a week? Compare their leaves.

## Explanation

Topsoil supports plant life because it contains material from living organisms that has decayed; leaves and branches fall to the ground and decompose along with such materials as animal droppings to produce soil *humus*, the rich black part of soil that nourishes plants. Plants grow readily in topsoil, but their growth is limited in soil below the Earth's surface, which contains very little decaying animal and plant matter.

## References

*Nature in a Nutshell for Kids* (both)
Jean Potter
John Wiley and Sons, Inc., New York, 1995

*The Kid's Nature Book* (children)
Susan Milord
Williamson Publishing, Charlotte, VT, 1989

## #5. Creating a Rockhound

**The Main Idea:** Rocks are everywhere: They are fun to touch and look at (especially the colorful ones), and some kids just love to throw them—especially into water! Children enjoy the different colors, shapes, and textures of rocks. In this activity, both of you can collect and learn about rocks and minerals, and in the process your child can learn how scientists classify information.

**Things You'll Need:**

- Empty egg cartons
- Old rags or towels
- Small hammer
- Goggles for adult and child
- Copper penny
- Glass marble

- Notebook
- Handheld magnifying glass

**How Long It Takes:** Optional

## What You'll Do

1. Collect rocks from your community or from a favorite park to fill at least 1 egg carton.
2. With your child far away, wrap a rag around one of your collected rocks and tap it with a hammer until it breaks into a few pieces. Avoid striking so hard that the sample is smashed.
3. For each sample, record in a notebook the shape of the rock, where it was found, how hard it was to break, the shapes of the pieces (jagged or smooth), color or colors, the pattern of colors (streaks of colors or just spots or blotches), and the texture (rough or smooth, dull or shiny). Use a magnifying glass to look closely at the details of the rock.
4. Test how hard the rock is by using your fingernails, a penny, and a glass marble. Can you scratch the rock with just your fingernail or with a penny? Can the rock scratch a penny and a glass marble?
5. Group rocks together that have similar characteristics. For young children, use a few characteristics of each rock to sort and classify them. Ask your child to pick a name for each group. On your next rock-hunting session, bring your notebook along and see whether you can add to each group.

| *Rock* | *Location* | *Color(s)* | *Pattern of Colors* | *Hardness* | *Texture* |
|---|---|---|---|---|---|
| 1 | ______ | ______ | ______ | ______ | ______ |
| 2 | ______ | ______ | ______ | ______ | ______ |
| 3 | ______ | ______ | ______ | ______ | ______ |
| 4 | ______ | ______ | ______ | ______ | ______ |
| 5 | ______ | ______ | ______ | ______ | ______ |
| 6 | ______ | ______ | ______ | ______ | ______ |
| 7 | ______ | ______ | ______ | ______ | ______ |

## Explanation

All rocks are made of minerals. Although there are about 2,000 different minerals, only about 100 are considered common. There are 3 basic classes of rocks: igneous, sedimentary, and metamorphic rocks. For older children especially, you can use a rock and mineral guide to identify each rock sample and its class.

*Igneous* rocks come from the cooling of magma, which is the hot, molten rock that exists below the Earth's crust and is released by volcanoes. Igneous rocks are relatively hard and evenly grained. Granite, which consists mostly of quartz, feldspar, and mica, is the most common igneous rock. *Sedimentary* rocks are formed from soil, pieces of shells, and pieces of rocks that are deposited by water. Sandstone and shale are examples. *Metamorphic* rocks are formed from the other two kinds of rocks that have been exposed to high temperatures and pressures for a long time. Quartzite, one of the hardest rocks, and marble are examples of metamorphic rocks.

## Digging Deeper: Name That Rock!

To help you identify some rocks, you may need to do some simple tests in addition to what is described in this activity. The back of an unglazed porcelain tile, for example, can be used for a streak test. Soft rocks like talc will leave a streak, while a hard one like quartz will not. When a few drops of vinegar are added to crushed limestone or marble, fizzing will occur (vinegar reacts with the calcium carbonate of the rocks to release carbon dioxide). There are many other tests that you and your child can learn about and use to identify rocks.

**It is important that you and your child follow some simple rules rock collecting. Rocks should always be kept out of mouths and eyes. Rock dust is an irritant, so thoroughly wash your hands after handling your rock samples. Some rocks—either before or after breaking—may have sharp edges, so handle them carefully or use gloves. When picking up a rock outdoors, you should wear gloves: An insect that stings or bites may be lurking underneath.**

## References

*The Audubon Society Field Guide to North American Rocks and Minerals* (adult)
Charles W. Chesterman
Alfred A. Knopf, 1979

*Rocks and Minerals* (both)
Herbert S. Zim and Paul R. Schaffer
Golden Press, 1957

*Geology* (both)
Frank H. T. Rhodes
Golden Press, 1972

## Web Site

*Ask-A-Geologist,* U.S. Geological Survey (adult)
URL: http://walrus.wr.usgs.gov:80/docs/ask-a-ge.html

# 36. Sky Watching

All of us, whether we live in crowded cities or in the wilderness, are sky watchers. Stars we see, and the patterns called *constellations* in particular, have attracted the attentions of both children and adults across the world for millennia. In the activities and experiments described here, your child can acquire some knowledge and understanding about what these heavenly bodies in the day and night skies are and how they affect our lives.

## #1. Starry, Starry Night

**The Main Idea:** By observing the night sky, you can learn much about stars, planets, the moon, and the Earth's rotation.

**Things You'll Need:**

- A clear night
- A place to stargaze away from lights
- A Blanket
- A Star chart
- A Flashlight

**How Long It Takes:** As short or as long as you like

## What You'll Do

1. Study a star chart with your child. Such charts can be readily obtained from libraries or bookstores or can be ordered from:

   The Astronomical Society of the Pacific
   1290 24th Ave.
   San Francisco, CA 94122

2. Lie down on a blanket, letting your eyes adjust to the dark.
3. Locate Polaris (the North star), the Little Dipper, and other constellations.
4. Notice that stars twinkle and have color. Are there planets that are visible?
5. If you stargaze for several hours, stop, and then return, notice that the constellations will have changed their positions in the sky. They appear to move from east to west because of the Earth's rotation.
6. Search for stars and constellations throughout the year, noting the differences in the skies as the seasons change. Make your own sky charts for each month.

## Explanation

The stars we see—like the most famous one, the sun—are extremely hot bodies that release light. The sun and almost all the stars we can see are in the Milky Way Galaxy, which is a huge system of billions of stars. Each night, as the Earth rotates toward the east, new stars and constellations become visible over the horizon while others disappear below the horizon to the west.

**Investigate the myths associated with some of the constellations, such as that of Hercules. Ask your child whether she sees different patterns of stars and what they look like to her. Perhaps she can think of her own myth surrounding the constellation.**

## #2. Constellation Viewers

**The Main Idea:** You can see the constellations on your ceiling or through small film canisters. By making these constellations and looking at them, you will be able to recognize them easily in the night sky. It is also fun for your child to copy the night sky on his bedroom ceiling.

**Things You'll Need:**

- Round empty cardboard box, such as one for oatmeal or potato chips
- Black construction paper
- Children's glue
- Flashlight
- A sharp point from a knife or scissors

**How Long It Takes:** Depends on the number of constellations you duplicate

## What You'll Do

1. Using an empty oatmeal or potato chip box, cut open the permanent side (the bottom).

2. Line the box and the removable side (the top) with black construction paper.
3. Cut out small openings in the removable top to mimic a constellation, as shown below.
4. At night, place a flashlight inside the tube and direct it toward the ceiling; then shut out the lights and turn on the flashlight. Your constellation will appear on the ceiling.
5. Make several constellations on the covers and see who can recognize them.

## Explanation

When you turn on the flashlight, the light will travel through the openings in the cover onto the ceiling. If you think of your ceiling as the night sky, you can project several of your favorite constellations.

You can also use the black canisters that contain 35 mm film to make constellations. Use a pin to poke holes where stars should be. Rotate the pin in order to make larger holes for stars that appear bigger in the night sky. Hold the canisters up to a light source, and you will see the pattern of the constellation. These canisters can be used at night to help you find the constellations in the sky.

## #3. When the Moon Comes over the Mountain

**The Main Idea:** A lunar month lasts about 28 days, during which time one full moon occurs. Over a month's time, you and your child can observe and demonstrate the moon phases.

**Things You'll Need:**

- A clear night
- Pencil and notebook
- Flashlight
- Ball
- Plastic or glass bottle

**How Long It Takes:** A few minutes each night for about one month

## What You'll Do

1. Each night observe the moon, starting with a full moon. Very infrequently, two full moons appear in a month, the second of which is called a "blue moon"—hence the saying "once in a blue moon."

2. In your notebook, sketch drawings of the moon.
3. Demonstrate the phases of the moon by placing the ball on the top of the bottle and shining a flashlight on the ball in a darkened room. By changing the position of the flashlight while both of you remain stationary, you will illuminate different amounts of the ball and simulate the phases of the moon. Be sure to demonstrate the moon's 4 phases: new moon, half-moon, full moon, and half-moon again.

## Explanation

The moon appears only because it is reflecting light from the sun. The moon changes each night because it is orbiting around the Earth as the Earth revolves around the sun. The phase of the moon, therefore, depends on the relative positions of the sun, Earth, and moon.

## #4. Rocks Around the Clock

**The Main Idea:** You can make a simple sundial with just rocks and a stick. This activity will help your child understand the relationship between the time of day and the position of the sun in the sky.

**Things You'll Need:**

- Rocks
- Stick or piece of wood
- Marker

**How Long It Takes:** A few minutes each hour

## What You'll Do

1. Insert a stick about $1\frac{1}{2}$ feet into the ground in a sunny, open area.
2. Place a rock each hour at the end of the stick's shadow.
3. Mark the time on the rock. For preschoolers, you can mark meal-times, or other times of the day that may be meaningful to him instead of the actual clock time.

## Explanation

Each day the sun rises in the east and travels across the sky to set in the west. A sundial uses the sun's shadow to tell time. The shadow appears at the opposite end of where the sun is as it moves across the sky. People thought for about 2,000 years that the sun was actually moving; we now know that the Earth is rotating on its axis so that the sun appears in the morning and then disappears at the end of the day.

For older children, you can construct a more elaborate sundial: Draw a circle on a posterboard or a piece of cardboard and insert a pencil or dowel in its center. Place the sundial in a sunny open area and write down the time each hour at the end of the pencil's shadow.

## References


*A Field Guide to Stars and Planets* (adult)
Donald H. Menzel and Jay M. Pasachoff
Houghton Mifflin Co., Boston, MA, 1983

*The Sky Observer's Guide* (both)
R. Newton et al
Golden Press, 1985

*The Night Sky Book: An Everyday Guide to Every Night* (children)
Jamie Jobb
Little, Brown and Company, 1977

*Sky All Around* (children)
Anna Grossnickle Hines
Clarion Books, New York, 1989

*The Glow in the Dark Night Sky Book* (both)
Clint Hatchett
Random House, New York, 1988

*The Heavenly Zoo: Legends and Tales of the Stars* (children)
Alison Laurie
Farrar, Straus and Giroux, New York, 1979

*Where Does the Day Go?* (children)
Walter M. Myers
Parent's Magazine Press, New York, 1969

*Sun Up, Sun Down* (children)
Gail Gibbons
Harcourt Brace Javanovich, New York, 1983

## Web Sites

*EXPLORANET* (both)
URL: http://www.exploratorium.edu

*NASA* (both)
URL: http://mosaic/arc.nasa.gov/nasaonline/nasaonline.html

*SEDS* (Students for the Exploration and Development of Space) (younger and older teens)
URL: http://seds.1pl.arizona.edu/

*Astronomy and Space on the Internet* (both)
URL: http://fly.hiwaay.net/-cwbol/astro.html

*The Nine Planets* (adult)
http://seds.1pl.arizona.edu/nineplanets/nineplanets/nineplanets.html

# 37. The Human Body

Kids are naturally curious about their own bodies and pose many questions about the body throughout their childhood. It is helpful for your child to learn about the human body early on, so that he or she can begin to understand the importance of forming healthy living habits.

## #1. A Taste of Honey

**The Main Idea:** Children love to use their senses in exploring the world. In this activity, you both can use your sense of taste to locate the different taste buds on your tongue. Then you can draw their locations, using symbols and words for sour, sweet, salty, and bitter areas.

**Things You'll Need:**

- Several cotton swabs
- Lemon juice
- Salt water
- Honey
- Coffee grounds

**How Long It Takes:** About 20 minutes

## What You'll Do

1. Draw a picture of your child's tongue. Dab a cotton swab with lemon juice and place the cotton on the different areas of the tongue until she can taste the lemon juice. Begin to fill in your drawing.
2. Do the same for each of the foods so that you can determine where the individual taste buds are located..

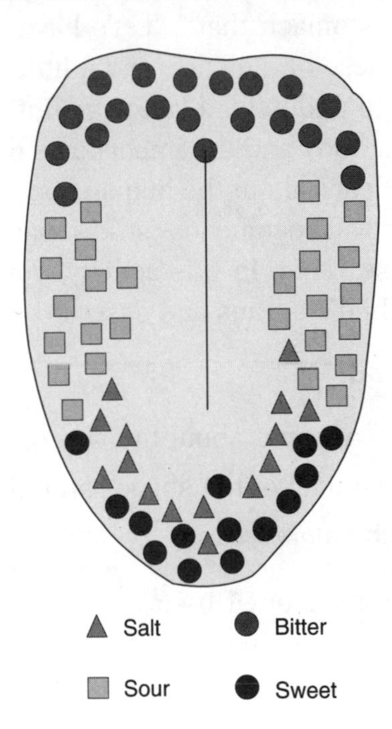


## Explanation

Our tongues contain about 3,000 taste buds; each has clusters of cells that send messages to your brain about what you are eating. There are 4 different tastes:

sweet, salty, bitter, and sour. The taste buds that can sense each taste are located in a specific area of the tongue, as shown in the illustration.

## #2. Major Organs of the Body

**The Main Idea:** Children hear names of organs all the time in everyday language: "I can't stomach that," "Let's have a heart to heart talk," etc. Learning where these organs are, and a little about what they do, helps children not only to understand language but also to appreciate the wonders of the human body and the importance of taking care of it. It is also helpful for kids to learn about the human body early, so that seeing health care professionals will become less stressful as they understand what will occur during these visits. In this activity, you and your child draw the location of your child's organs and describe a little about what they do.

**Things You'll Need:**

- 2 large pieces of paper about the size of your child
- A diagram or model of the shapes and location of organs
- 8 $\frac{1}{2}$-by-11-inch colored paper

**How Long It Takes:** About 1 hour

## What You'll Do

1. Lay the 2 pieces of paper beside each other on the floor. Ask your child to lay down on each paper and trace her body onto the paper.

2. Use the sheets of colored paper to make pictures of organs and then cut out the pictures. Use different colored sheets so that each organ stands out.
3. Read about the functions of the different organs and write their names, along with what they do, in your child's words on the cutouts. For preschoolers, you can draw pictures showing their functions in the body.
4. You and your child can then proceed to place the organs on the outline of your child so that he can see exactly where they are located. After a while, ask him to place the organs in their appropriate locations by himself.

## Explanation

Use children's books to learn about the human body, being careful to avoid graphic photos that your child may find frightening. Some suggested books are listed at the end of this section.

**You may enjoy doing this activity each year, charting your child's growth and also helping your child to learn more about the human body. If you have a pet, you can draw your pet's outline on a large sheet of paper and again attach organs in their appropriate locations. This activity shows how similar animals are to humans and how animals also need good nutrition and exercise.**

## #3. Can't You Hear My Heart Beat?

**The Main Idea:** Here is a simple way to make a stethoscope that your child can use to listen to her heartbeat—and yours as well. This activity helps demonstrate that our hearts are like pumps, pushing blood throughout our bodies.

**Things You'll Need:**

- Small funnel
- Glass or plastic T- or Y-tube
- Rubber or tygon tubing

**How Long It Takes:** 30 minutes

## What You'll Do

1. Place a short piece of rubber or tygon tubing on the stem of a small funnel. Small funnels are usually available in houseware departments or hardware stores. Tubing can also be found in most hardware stores. Attach the other end of the short piece of tubing to the end of a glass or plastic T- or Y-tube. (T- or Y- tubes can also be found in most hardware stores.)
2. Attach some 4- or 5-inch-long pieces of tubing to the other 2 branches of the T- or Y-tube. If available, you can also attach 2 small funnels to the other ends of these longer sections of tubing for easier listening.
3. To use this stethoscope, hold the funnel firmly over your child's heart while he holds the ends of the long tubes in his ears. (It may be easier for him if you attach the small funnels suggested in step 2.) You could

also hold the funnel against your own heart for him and ask him whether it beats slower or faster than his own.

4. He could also listen to his heart after exercising; the heart must pump faster to bring nutrients and oxygen to working muscles.
5. Listen to heartbeats from different points on your chest. Do they get softer as you move from the left to the right side of your chest?

## Explanation

The human heart is located to the left of the center of the chest and is pointed to the right. It is a strong muscle that beats continuously throughout life, pumping blood around the body so that all cells receive nutrients and oxygen. An adult heart beats about 70 times each minute at rest, while a child's heart beats much faster—between 90 to 120 beats per minute.

## Digging Deeper: Pulse Rate and Exercise

You can also hear your heart pump at your wrist: Firmly place 2 fingers on the inside of your wrist with your thumb pushing against the other side of your wrist. After your child has found her pulse, ask her to begin counting until you say stop. You can do this for 15 seconds and then multiply by 4 to obtain the pulse rate. She can measure your pulse rate as well and then compare. You can both then exercise and measure your pulse rates. Construct a table in which you record rates at rest and after walking and running.

| | *At Rest* | *Walking* | *Running* |
|---|---|---|---|
| Parent | ________ | ________ | ________ |
| Child | ________ | ________ | ________ |

## References

*The Human Body and How It Works* (children)
Angela Royston
Random House, New York, 1991

*Blood and Guts, A Working Guide to Your Own Insides* (children)
Linda Allison
Little, Brown and Company, Boston, 1976

*Human Anatomy for Children*
Ilse Goldsmith
Dover Publications, Inc., New York, 1964

*What's Inside My Body* (children)
Angela Royston
Dorling Kindersley, New York, 1991

*The Human Body at Your Fingertips* (children)
Judy Nayer
McClanahan Book Co., 1995

*My First Body Book* (children)
Lara Tankel Hotz, Editor
DK Publishing, Inc., New York, 1995

# 38. Weather

Weather is with us every day—whether we like it or not! It's a topic that children (and many parents) are generally very curious about. You and your child can become amateur meteorologists, monitoring and trying to predict the weather with a homemade weather station; or you can do some of the simple, easy experiments below that will help your child understand weather events. Some children become less fearful of events like lightning and thunderstorms when they understand their causes.

## #1. Raindrops, So Many Raindrops

**The Main Idea:** In this simple and quick activity, your child can see how raindrops form in clouds.

**Things You'll Need:**

- Glass jar with a cover
- Ice cubes

**How Long It Takes:** Just a few minutes

## What You'll Do

1. Fill the jar with ice cubes and cover.
2. After a few minutes, touch the outside of the jar. Does it feel wet? Can you see raindrops forming on the outside surface of the jar?

## Explanation

Water in the form of a gas (water vapor) is always in the air. The cold jar cools the surrounding air, causing some of the water vapor in it to condense and form liquid water on the outside surface of the jar. Raindrops form in clouds in the same way: Warm air from the ground rises and is cooled by the cold air above, causing drops of water to form clouds. When the drops get larger, they fall as rain.

## Digging Deeper: The Big Chill

Form raindrops inside the jar by filling it halfway with warm tap water. Use a funnel so you do not get water on the sides of the jar as you are pouring it. Cover the jar and place it in the freezer; observe the jar every 5 minutes or so. Do you see condensation drops forming on the inside wall of the jar? The air inside the jar is much warmer than the air in the freezer; when the air in the jar cools, water vapor condenses on the inside surface.

Condensation of water vapor occurs all around us: at restaurants on the outside of drinking glasses filled with ice water, on the inside of our windows during the winter, on our eyeglasses when we venture from the air-conditioned indoors out into the blistering heat during the summer, etc. Discuss this with your child whenever the opportunity arises. Remember, science is always all around us!

## #2. The Pressure of Building a Barometer

**The Main Idea:** A barometer can be used to help your child understand air pressure and can also help in predicting the weather.

**Things You'll Need:**

- Glass jar
- Tall plastic or glass bottle
- Large balloon
- Straw
- Scissors
- Rubber bands
- Tape
- Paper label
- Books

**How Long It Takes:** A few minutes to build

## What You'll Do

1. Cut the balloon and use it to cover the glass jar, securing it tightly with rubber bands.
2. Cut the end of the straw into a point and tape it to the top of the jar (see below).
3. Place a large paper label on the glass or plastic bottle. (If you are using a plastic bottle, place some rocks inside to keep it stable.)
4. Line up the bottle with the straw so that the straw end is on the label. Mark where it is. Place books on both sides of your setup to keep it stable.
5. Each day note the position of the straw pointer. Is it higher or lower? Is it higher or lower on rainy days? On sunny days?

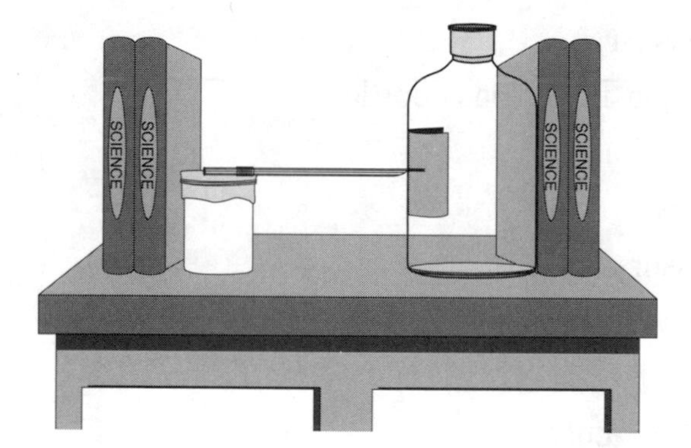


## Explanation

The air pressure, or atmospheric pressure, is how hard the air presses down on any one place on Earth. As air pressure rises, or increases, the balloon is pushed down, and the straw pointer tilts upward. When the air pressure falls, the balloon rises, and the pointer slants downward. Generally, low pressures predict storms, and high pressures indicate fair weather.

With older children especially, you can compare your barometer with the pressure reported by your local weather station. Write the actual pressure (usually in inches or millimeters of mercury) on the label next to the pointer of your barometer. After several readings are marked on your barometer, you can estimate pressures yourself.

## #3. Raindrops Keep Fallin' in My Rain Gauge

**The Main Idea:** The amount of rainfall associated with individual storms, or over a month or season, is another element of weather that amateur meteorologists can measure. You can measure how much rain falls in your neighborhood.

**Things You'll Need:**

- Clear glass jar
- Tape
- Marking pen

**How Long It Takes:** A few seconds

## What You'll Do

1. Place some tape vertically on a jar with straight sides. With older children, copy a ruler that has both inches and centimeters.
2. Leave your rain gauge in an open area where rain will not be obstructed by buildings or trees.
3. After it rains, measure the inches and centimeters of rainfall from your gauge and record the amount. Compare with weather reports.

**Measuring rainfall is a good opportunity to use the metric system with your child. Science throughout the world relies on the metric system; the earlier it is introduced, the easier it will be for your child to use the units in the system later on.**

## #4. I've Looked at Clouds

**The Main Idea:** By watching and drawing clouds over a period of time, you can learn what they are and how they can often help you predict the weather. Children naturally tend to notice their different shapes and forms, and enjoy watching clouds drifting by.

**Things You'll Need:**

- Cotton balls
- Children's glue
- Blue paper
- Gray or black paper

**How Long It Takes:** About 10 minutes each day to make pictures of that day's cloud formations. If you make this activity a long-term project, your child will be able to see the relationship of cloud patterns to the weather.

## What You'll Do

1. Observe the cloud formations for several days and sketch their shapes on a blue piece of paper that resembles the sky. For preschoolers especially, you can use cotton balls to fill in the shapes and thereby duplicate more accurately the way the clouds looked.
2. Record in a notebook the overall pattern of clouds as well as their shapes: Were the clouds in puffy layers or did they appear in patches? Use your child's description of the way the clouds look to him. Also record the weather for each day so that you will be able to see a relationship between cloud formations and weather patterns.
3. After observing clouds for a while, you and your child can try using cloud formations to predict the weather.

## Explanation

Clouds play a major role in all of our lives: Without them there would be no rain or snow, no thunder and lightning, and no rainbows. Not only would the weather be monotonous, but so would the sky—without the endless different shapes and forms that clouds take. Clouds are visible collections of tiny water drops or ice crystals that are in the air. Clouds can be thick or thin; they can be high, middle, or low in the sky; and they can have a large variety of different shapes that form different patterns. Clouds can be classified in 10 basic groups that are helpful in weather prediction. Some examples include small puffy *cumulus* clouds that are a sign of fair weather; low, uniform grayish *stratus* clouds from which drizzle may fall; and wispy *cirrus* clouds that signal an approaching warm front that may bring rain or snow.

## Digging Deeper: Identify Cloud Types

For older children especially, you can explore the 10 basic cloud types and identify each day's formation. Most weather or meteorology books contain descriptions along with photos and pictures of the different cloud types and their relationships to impending weather.

## References

*Weather Watch* (children)
Valerie Wyatt
Addison-Wesley, Reading, MA, 1990

*Bob The Snowman* (children)
Sylvia Loretan
Viking, New York, 1991

*Raindrop Stories* (children)
Preston Bassett with Margaret Bartlett
Four Winds Press, New York, 1981

*Earth Child* (both)
Kathryn Sheehan and Mary Waidner
Council Oak Books, Tulsa, OK, 1994

*Weather* (both)
Paul E. Lehr
Golden Press, 1965

*Nature's Weather Forecasters* (children)
Helen R. Sattler
Thomas Nelson, Inc. Publishers, 1978

## Web Sites

*Weather Processor* (adult)
URL: http://thunder.atms.purdue.edu

*Penn State University Weather Pages* (adult)
URL: http://www.ems.psu.edu/wx/index.html

*National Hurricane Center* (both)
URL: http://nhc-hp3.nhc.noaa.gov

## SECTION 4

# Games and Puzzles

Playing games, solving puzzles, collecting leaves, and doing other similar activities together is one of the best ways to enjoy your child and, at the same time, encourage her interest in science and nature. Needless to say, children love to play, and when you join in, you can make these moments very special. Many of these activities can also engage your child during trying times, such as at the airport, waiting for delayed flights; in restaurants with slow service; or even in long lines that never seem to move.

There are many games, puzzles, and other activities that will address the interests and skills of your youngster and entertain her as well. Some of the things young children enjoy include make-believe, sorting, identification, and guessing games, as well as writing projects and cooperative projects like planting a garden. Grade school kids especially like simple games with rules.

# Guessing Games

Guessing games can build memory and problem-solving skills.

## 39. The Name Game

**The Main Idea:** Each of you can think of an animal or plant and then take turns trying to guess its identity.

**Things You'll Need:**

- Information about different animals or plants

**How Long It Takes:** Optional. (It may depend on the length of the car trip or line, for example.)

### What You'll Do

1. Read some books with your child about different animals or retrieve some information from the Internet. Many children are intrigued by animals; some prefer to learn about large ocean mammals like whales, while others prefer cuddly koala bears. Find out your child's preferences and then together learn all you can about where they live; what they eat; how they walk; who their enemies are; and, of course, what they look like.

2. Armed with this information, you can ask your child to try to guess the animal you're thinking of. Ask questions like these: How big is it? Does it live in the water or on land? Does it live near us? Give hints along the way—especially if he seems perplexed. You, of course, can play the same guessing game with plants and trees in your community.

**There are various ways to modify the game; for example, you can ask your child to think of 4 animals that have 4 legs and graze, or 5 animals that live in Africa and are carnivores. Encourage her to make up the next category. The possibilities are endless—and can get you through waiting in a very long line. The more information you obtain together, the more detailed these games can be. This type of game will also help your child see similar characteristics among living things, which is part of what biologists do when they classify animals and plants.**

## References


*Planting a Rainbow* (children)
Lois Ehlert
Harcourt Brace Jovanovich, New York, 1988

*Animals That Live in Trees* (both)
Jane R. McCauley
National Geographic Society, Washington, D.C., 1986

## Digging Deeper: Endangered Species

Sometimes you can limit your guessing game to endangered species. This a way of teaching your child about the needs of animals and how they can disappear if we humans don't learn to share resources with them. Most species that are endangered are suffering from habitat loss. The Nature Conservancy publishes a list of the most endangered 500 plants and animals in the United States today. Obtain information and let the games begin!

*The Modern Ark: The Endangered Wildlife of our Planet* (children)
Claire Littlejohn
Dial Books for Young Readers, New York, 1989

*Animals in Danger* (children)
Janine Amos
Raintree Steck-Vaughn Publishing, 1993

# 40. The Who

## What You'll Do

Children often imitate animals by pretending to eat out of bowls on the floor like the family dog or cat, or hanging from trees like a chimpanzee. Encourage your child to act like any animal he chooses and then try to guess what he is. (This activity is not recommended for most public places.) Obtain some information on the behaviors and activities of different animals, so that he will have a wide range of behaviors that he can draw from to make the game more interesting.

## Reference

*Nature for the Very Young* (both)
Marcia Bowden
John Wiley and Sons, New York, 1989

# 41. A Chipmunk for a Day

## What You'll Do

Ask your child to select a favorite animal and pretend the animal has a diary. Together, write an entry into the diary for one day. For a chipmunk, for example, it might include digging a burrow for shelter, collecting leaves and twigs for her home, burying acorns and walnuts for the winter, a narrow escape from an owl or dog, etc. It is also important to imagine how the chipmunk felt as she ran from the dog and how tired she was at the end of the day. Many children are unaware of the fact that animals experience feelings but "if you walk the footsteps of a stranger, you'll learn things you never knew you never knew."

## References

*Wild Fox* (children)
Cherie Mason
Downeast Books, Camden, ME, 1993

*A Wolf Story* (children)
David McPhail
Charles Scribner's Sons, New York, NY 1981

*Little Mop Lost* (children)
Kayoko Kanome
Carilorhoda Books, Inc., Minneapolis, MN, 1992

*Helping Our Animal Friends* (both)
Judith E. Rinard
National Geographic Society, Washington, D.C., 1985

*Every Living Thing* (children)
Cynthia Rylant
Bradbury Press, 1985

# The Puzzle Place

This is another activity in which you can join forces with your child, learn about any scientific topic you choose, and turn your home into a puzzle place.

## 42. Jigsaw Puzzles for Fun and Profit

**The Main Idea:** In this activity, you and your child can combine your artistic talents and your knowledge of nature to create jigsaw puzzles that your child can play with for days later, remembering the fun he had making them with you.

**Things You'll Need:**

- Cardboard or posterboard
- Nontoxic, lead-free markers or crayons (some foreign-made brands may contain lead)
- Scissors
- A ruler

**How Long It Takes:** For preschoolers who use larger pieces, only about 10 minutes. Because older children will draw more complex pictures and will need smaller pieces, more time will be needed.

## What You'll Do

1. Decide on a topic that interests your child, such as animals, plants, stars, etc. Draw on a piece of cardboard a habitat, perhaps a forest scene with animals and plants. Don't forget to include the critical parts of a habitat, like food sources, water, and shelter.
2. Turn over the drawing and mark out the shapes of the pieces. For a very young preschooler you could use squares, triangles, and rectangles to help her learn the different shapes.
3. Cut out the pieces and then give it a trial run with your child to make sure they fit.
4. You could also use photographs or copies of pictures from books by gluing photos and illustrations onto a piece of cardboard.

**These handmade jigsaw puzzles can also be presents for your child to give to her friends or relatives on special occasions.**

# #43. Personal Crossword Puzzles

No cross words—just crossword puzzles! Agree on a theme that interests your child and create your own crossword puzzle. For example, the crossword puzzle below is for a youngster who likes to watch birds high up in a tree.

## BIRD THINGS

| 1. | | | 2. | | 3. | | |
|---|---|---|---|---|---|---|---|
| | | | | | | | |
| | | 4. | | | | | |
| | | | | | | | |
| 5. | | 6. | | | 7. | | 8. |
| | | 9. | | 10. | | | |
| | | | | | | | |
| 11. | | | | | 12. | | |

**ACROSS**

1. WHAT A BIRD LAYS HER EGGS IN.
3. WHAT A BIRDLING HATCHES FROM.
4. ARE USED TO BUILD NESTS.
5. ____ THE BIRDS.
7. WHAT BIRDS DO AT YOUR FEEDER.
9. BIRDS USED THESE TO GRASP ON TO BRANCHES.
11. MANY BIRDS START FEEDING AT_____.
12. WHAT PEOPLE THINK BATS DON'T DO WELL.

**DOWN**

1. _____HATCH.
2. A RUFOUS SIDED_____
3. A YOUNG EAGLE.
5. YOU PUT THESE IN YOUR BIRDFEEDER.
6. FIRST LETTERS OF TWO MEAT-EATING LARGE BIRDS.
7. EGR_____.
8. MANY BIRDS BUILD THEIR NESTS IN A_____
10. ROAD RUNN_____.

# SOLUTION

| 1. N | E | S | 2. T | | 3. E | G | G |
|---|---|---|---|---|---|---|---|
| U | | | O | | A | | |
| T | | 4. T | W | I | G | S | |
| | | | H | | L | | |
| 5. S | A | 6. V | E | | 7. E | A | 8. T |
| E | | 9. F | E | 10. E | T | | R |
| E | | | | R | | | E |
| 11. D | A | W | N | | 12. S | E | E |

# 44. The Seekers

Again, agree on a theme and then hide in boxes relevant words that your child will try to locate. Depending on the age of your child, you can arrange words upside down, right to left, or diagonally. Below is an example using astronomy as the subject pertaining to the hidden words.

**FIND THE HIDDEN ASTRONOMICAL WORDS**

| S | U | N | E | P | T | U | N | E |
|---|---|---|---|---|---|---|---|---|
| K | T | E | B | V | E | N | U | S |
| Y | O | A | C | I | O | D | K | O |
| B | E | A | R | L | G | I | P | L |
| C | O | M | E | T | I | P | U | A |
| L | E | O | A | P | C | P | R | R |
| E | D | O | G | R | Y | E | S | T |
| O | I | N | L | A | S | R | A | E |

(WORDS APPEAR RIGHT TO LEFT, TOP TO BOTTOM, AND DIAGONALLY ONLY.)

# 45. Moving On

Create a maze that pertains to animals traveling to locations where food is available. Some ideas include a grizzly bear in Alaska on her way to a river containing salmon, a brown bear looking for honey and lost in a maze of forest trails, a herd of elk seeking vegetation during the winter, or an Indigo bunting migrating north. Here is an example of a maze in which you can help elk find their way to greener pastures.

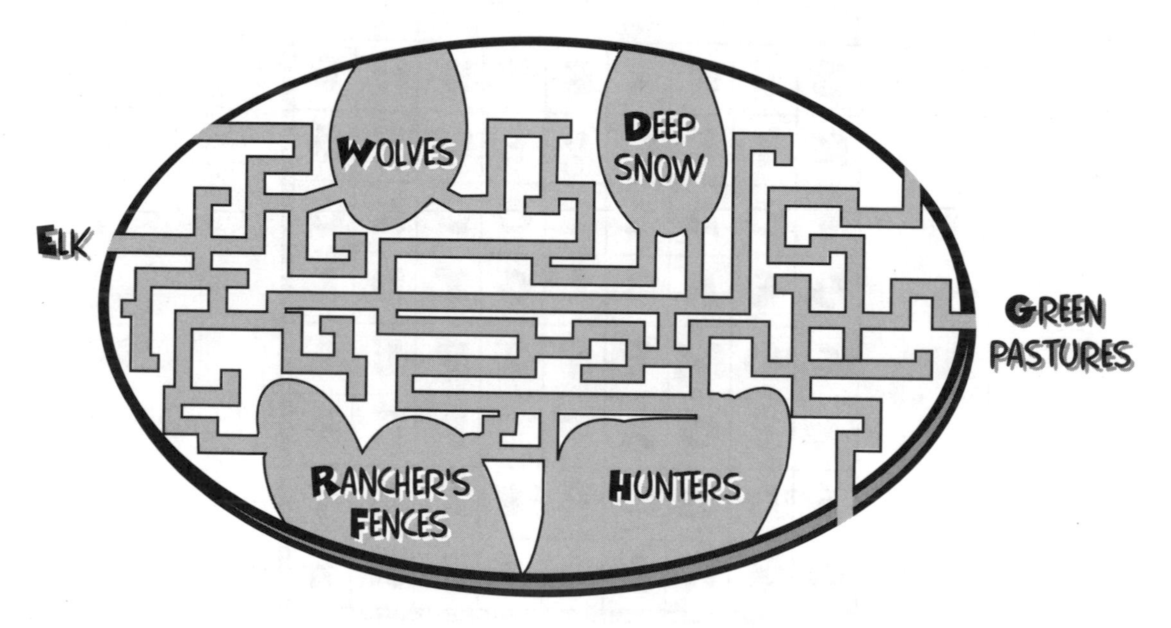

# The Games Children Play

Below are some other games that you can play to engage your youngster's interest in science and nature.

## 46. Board Games

Six- to eight-year-olds especially enjoy playing simple games with rules. Try making up a game of your own that focuses on your child's particular interests. Below is an example that centers on the difficult journeys of migrating birds. You can make the game more relevant to your child if you select a migrating bird in your community and find out the bird's final destination.

START
CLEAR WEATHER MOVE TWO MORE SPACES.
YOU CAN'T FIND A PLACE TO STOP. LOSE YOUR NEXT TURN.
GO TO THE POND.
A STORM IS APPROACHING GO BACK TWO SPACES.
GO TO THE FOREST.
STOP AT A BIRDFEEDER AND TAKE ANOTHER TURN.
POND
REST, EAT, AND DRINK. THEN MOVE TWO SPACESS.
AVOID AN OWL AND STAY HIDDEN. LOSE YOUR NEXT TURN.
TAKE COVER FROM A SNOW STORM IN A PINE TREE.
FOREST
EAT, SLEEP, AND REST. THEN TAKE ANOTHER TURN.
A WINDY DAY SLOWS YOU DOWN. GO BACK TWO SPACES.
GO TO THE BACKYARD HABITAT.
HIDE IN A TREE TO AVOID A HAWK OVERHEAD.
WINDS AGAINST YOU. MOVE BACK ONE SPACE.
WINDS BEHIND YOU. MOVE THREE SPACES FORWARD.
APPROACHING THE LAST LEG OF YOUR JOURNEY.
GO TO THE TALL GRASS.
IN THE HOME STRETCH.
BACKYARD HABITAT
REST, EAT, THEN TAKE ANOTHER TURN.
EXHAUSTED STOP AND REST. LOSE ONE TURN.
GETTING CLOSER.
TALL GRASS
REST AND SEARCH FOR FOOD AND WATER
FINAL DESTINATION
MIGRATING BIRDS
EACH OF YOU CAN DRAW A BIRD ON CARDBOARD AND CUT IT OUT TO PLAY THE GAME. USE ONE DIE ONLY AND BEGIN YOUR JOURNEY.

# 47. Scavenger Hunts

## What You'll Do

Make a list of items to collect during a nature hike through a forest or along a beach, for example. With younger children, you can search for red birds, pine cones, acorns, daisies, clam shells, or sand dollars.

## Digging Deeper: The Search for Beanie Babies

For older children, you can devise a more sophisticated scavenger hunt that will challenge their powers of observation as well as their knowledge. The idea is for them to locate notes that you hide which lead eventually to an endpoint, such as a bed of wildflowers or one of the Beanie Babies®. You can start the children off with a note like "Find the large white oak tree surrounded by poplar and yellow pine trees, and turn over the rock that lies at its base." Be specific so that you tap into your child's knowledge of the natural world.

# 48. Computer Games

There are many different interactive programs on science and nature topics in the form of software and CD-ROMs that will help your child learn scientific skills. A few of these are listed here:

*Bumptz Science Carnival*
1-800-955-8749
URL: http://www.theatrix.com

*Junior Nature Guide Series: Insects*
1-416-868-6423
URL: http://www.natureguides.com

*Magic School Bus Explores Inside the Earth*
1-800-426-9400
URL: http://place2.scholastic.com/magicschoolbus/index.htm

To find effective products that your child will enjoy, you can subscribe to:

Children's Software Revue
44 Main Street
Fleminton, NJ 08822
URL: http://www.childrenssoftware.com

For older children, you can use the Internet for online scavenger hunts. Together you can make a list of facts to find relating to a favorite topic. For example, if your child enjoys astronomy, use the Internet to find the difference between a comet and a meteor, how many moons Saturn has, which planet is the smallest, etc.

## SECTION 5

# Places to Go, Things to Do

This final section is filled with suggestions for special places to visit with your child, along with tips on how you can make your trips enjoyable, educational, and memorable. Also included are suggestions for activities that highlight seasonal events. By making your connection with nature and science a daily activity, and by encouraging your child's curiosity about the natural world, you can help your child develop a love and appreciation for science and the environment that can last a lifetime.

# 49. Visiting Special Places

There are many different places that you can visit with your children to nurture their interest in science and nature. They might include natural environments, such as a pond or produce farm; community services, such as a water treatment plant; or elaborate hands-on science museums, such as the Exploratorium in San Francisco. No matter where you live, there are places to visit that can stimulate your child's sense of wonder about the world. Even a trip to a grocery store can be an interesting experience: You could discuss why some foods are kept frozen and others refrigerated, why fruits and vegetables should be thoroughly washed, or how the checkout scanner works. By seeing science at work in neighborhood places, your child will learn that science and nature are not only in laboratories, museums, botanical gardens, and zoos, but are also part of our everyday lives.

## Natural Places

A pond, creek, park, beach, river, or tree can serve as a special place where you and your child can share some of the mysteries of nature. If you visit a pond, find out about the plants, fish, amphibians, and animals that live or visit there: How do they adjust to the change in seasons? What evidence do they leave (tracks, eaten leaves, etc.)? What happens during a drought or periods of heavy rainfall? Study the pond itself: What materials get into the water? Where does its water come from (springs or runoff)? Is the pond being affected by people?

Keep a journal on your special place, noting its characteristics and changes. If you have selected a tree, observe its changes with the passing seasons and think about the reasons for these changes: Why does the tree lose its leaves during the fall? How does it spread its seeds? What birds visit or build nests in the tree? How deep are the roots? Look for seedlings. The more you observe, the more questions will be raised, which will foster your child's curiosity and lead to more discoveries about the natural world. It's not the amount of information that your child acquires that's important, but the desire to invest the time and energy to continue to learn and think about topics in science and nature.

## Your Community

Look for factories, industries, manufacturing plants, farms, businesses, or government services that can provide an opportunity to think about and discuss some scientific topics. For example, how do dry cleaning stores clean clothes? What chemicals do they use? How do they dispose of these chemicals? Do you think it's a good idea to let your dry-cleaned clothes air out before you wear them? Many industries and services conduct tours of their buildings for the public that your child might find interesting. Where does your water come from and where does it go? How about garbage and recycled materials—where do they end up? Visit these public services so that your child can see how scientific principles have been applied to make their lives more comfortable.

## Family Vacations

Vacations can serve as wonderful scientific expeditions for your child. If you're traveling to another state or country, compose a list of questions to be answered, such as: What plants grow here? Why are they different from those at home? Do the same trees grow here as at home? Do the constellations appear in the same place in the night sky? Is the climate different? Visit local museums, zoos, and botanical gardens. What exhibits relate to the region? Vacations can be an exciting way to appreciate the great diversity of life that flourishes on our planet.

## Field Trips

Field trips can be instrumental in nurturing your child's interest in the world around him, as well as helping him develop the observational skills that are so important in science. Field trips can include a walk in a park, a trip to the shore, a visit to a lake or pond, or just exploring your backyard. All these places provide numerous opportunities to learn about science and nature. Decide in advance with your child what you will be doing on your field trip, and bring your science and nature notebook to write down your activities. Will you be identifying different trees in a park or searching for wildflowers? Compare the texture of the different barks of trees, their leaf shapes and colors, their heights, and other attributes to identify the trees. Help your child look carefully and observe the finer details.

## Museums, Planetariums, Zoos, Nature Centers, and Botanical Gardens

Today, many of these special places—science and natural history museums, zoos, nature centers, and botanical gardens—are being designed to appeal to both children and parents. Some museums are completely oriented toward the learning needs of older children, while others contain discovery or hands-on rooms for toddlers and children in elementary school. Children's museums in particular offer exhibits that are appropriate for the developmental requirements of kids of different ages. These exploratory-type museums enable kids and parents to learn by doing. Zoos, nature centers, and botanical gardens are also developing into more of an interactive experience for visitors. Look for these special places when you plan your trips.

To help make your visits positive experiences for you and your child, keep in mind a few simple tips:

## Let your child lead the way.

Remember that it is not necessary to see the whole museum or nature center in one visit; what is important is that your child is stimulated by the environment and is enjoying the experience. Some kids are so fascinated by an exhibit that they remain and interact with it for hours. (So much for the idea that kids have short attention spans!) Let your child select what he wants to see.

## Discuss the special place before you visit.

Find out about the special place before you visit and talk to your child about what will be there. Most likely she will express some preferences almost immediately about what she would like to see and do. Call either the institution itself or the chamber of commerce or visitor's center for that region for information.

## Let your child participate in the exhibits in her own way.

People learn in different ways: Some are visual and learn more from written material, others benefit more from verbal discussions of what they observe, and others learn by touching and doing. Give your child space to approach the exhibits in her own way.

# 50. The Four Seasons

The four seasons offer such a wide a variety of unique and special activities—be sure to take advantage of them with your child. Here are a few suggestions to help you get started.

## Winter

Catch snowflakes on black paper and examine them with a magnifying glass; follow animal tracks; locate your favorite constellations during the winter; look for birds' nests and try to identify the kind of birds that made them; blow bubbles outside in the cold and compare them to bubbles blown indoors.

## Spring

Collect wild flowers; put out string, thread and any materials that birds can use to build their nests, and then look for them during the winter; cover a crocus with a box for several minutes and then look at how it closed its petals; look for earthworms (especially when it rains), observe, and study them; obtain some frogs' eggs from a pond or supplier and watch them become tadpoles and frogs (don't forget to return them to the pond); look for salamanders; start a compost bin; start planting a garden so that you can make vegetable soup from your produce; go on a nature hike collecting seeds; fly a kite (the winds of March are great for this).

## Summer

Obtain some pond water and observe it with a magnifying glass to identify what is in it; catch—for just a little while—fireflies (lightning bugs) in a clear jar and watch them with a magnifying glass to follow their blinking patterns; examine a flower and identify the different parts; record the sounds of animals at night and then try to identify them; look for your favorite constellations; look for insects under rocks, in grasses and on trees; listen to bird songs and identify their sources.

## Autumn

Collect leaves (you guessed it); collect wildflower seeds; check out the constellations; make a bird feeder to help the birds survive the winter; examine a pine cone to look for seeds; look for different tree seeds (how do they disperse?); find an empty bird's nest and carefully pull the nest apart with gloves, examining the materials it is made of; compost leaves and observe their decomposition.